When Old Technologies Were New

620
M

5-24-90

When
Old Technologies
Were New

Thinking About Electric
Communication in the Late
Nineteenth Century

Carolyn Marvin

OXFORD UNIVERSITY PRESS

New York   *Oxford*

Oxford University Press

Oxford   New York   Toronto
Delhi   Bombay   Calcutta   Madras   Karachi
Petaling Jaya   Singapore   Hong Kong   Tokyo
Nairobi   Dar es Salaam   Cape Town
Melbourne   Auckland

and associated companies in
Berlin   Ibadan

First published in 1988 by Oxford University Press, Inc.,
198 Madison Avenue, New York, New York 10016-4314

First issued as an Oxford University Press paperback, 1990
Oxford is a registered trademark of Oxford University Press

Library of Congress Cataloging-in-Publication Data
Marvin, Carolyn.
When old technologies were new.
Includes index.
1. Telecommunication—History—19th century.
2. Electric lighting—History—19th century. I. Title
TK5102.2.M37   1987   621.38   86-33339
ISBN 0-19-504468-1
ISBN 0-19-506341-4 (PBK)

Some parts of Chapter Four, "Dazzling the Multitudes:
Original Media Spectacles" are based on Carolyn Marvin,
"Dazzling the Multitude: Imagining the Electric Light
as a Communications Medium," in Joseph J. Corn, ed.,
Imagining Tomorrow: History, Technology, and the American Future
(Cambridge, Mass.: MIT Press, 1986).

9   8   7   6   5   4

Printed in the United States of America
on acid-free paper

For
Jean H. Marvin
Conrad W. Marvin

Acknowledgments

I wish to thank Lord Asa Briggs, whose interest in predictions about wireless technology sparked my own interest in the social beginnings of electric communications technology when I worked for him as a research assistant at the University of Sussex many years ago. The University of Pennsylvania provided summer research support at an important point in my work, and my colleagues at Penn in the Mellon Seminar on Technology and Culture gave an especially attentive and helpful reading to portions of this study that I presented to them. Throughout I have been grateful for the atmosphere of warm intellectual support at The Annenberg School of Communications from colleagues who take little for granted, but whose shared commitment to the investigation of communication as a *social* phenomenon kept me moving in the direction of the key ideas from which the theoretical structure of this study emerged. I am grateful to my research assistants at the Annenberg School, who cheerfully executed an endless number of assignments. I owe a special debt to Gwenyth Jackaway, who helped from afar, and to Pam Inglesby, who lent immeasurable assistance, and always with the greatest intelligence and care. Janice Fisher was a virtual rock of editorial aid and good judgment, and Gilda Abramowitz's valuable suggestions made this a more readable and precise piece of prose than it otherwise would have been. To Christine Bachen, Mary Mander, and Shari Robertson go my deepest gratitude. They listened to everything, and always gave the best advice.

May 1987   C.M.

Contents

Illustrations follow page 108

When Old Technologies Were New

This is the simmering of the electrical imagination, which fairly bubbles in the notion that in time the race may develop a special electrical sense.

—Park Benjamin

For media historians, the phenomenon of twentieth-century electronic mass media lies like a great whale across the terrain of our intellectual concern. Asked to explain what sort of phenomenon it is, most of us will unhesitatingly point to the hundreds of millions of radio and television sets that are bought by consumers and promoted by vast industries. This artifactual notion is pervasive and not much debated, for it seems simple, obvious, and convenient. But it has rendered invisible important aspects of electric media history, and perhaps of mediated communication generally. It does this in part by fixing the social origin of electric media history at the point when media producers began to service and encourage the appliance-buying demand of mass audiences. Everything before this artifactual moment is classified as technical prehistory, a neutral boundary at which inventors and technicians with no other agenda of much interest assembled equipment that exerted negligible social impact until the rise of network broadcasting. But a great deal more was going on in the late nineteenth century. New electric media were sources of endless fascination and fear, and provided constant fodder for social experimentation. All debates about electronic media in the twentieth century begin here, in fact. For if it is the case, as it is fashionable to assert, that media give shape to the imaginative boundaries of modern communities, then the introduction of new media is a special historical occasion when patterns anchored in older media that have provided the stable currency of social exchange are reexamined, challenged, and defended.

The present study is not, therefore, an effort merely to extend the traditional time line of electric media. It introduces issues that may be overlooked when the social history of these media is framed exclusively by the instrument-centered perspective that governs its conventional starting point. It argues that the early history of electric media is less the evolution of technical efficiencies in communication than a series of arenas for negotiating issues crucial to the conduct of social life; among them, who is inside and outside, who may speak, who may not, and who has authority and may be believed. Changes in the speed, capacity, and performance of communications devices tell us little about these questions. At best, they provide a cover of functional meanings beneath which social meanings can elaborate themselves undisturbed.

If artifactual approaches foster the belief that social processes connected to media logically and historically begin with the instrument, then new media are presumed to fashion new social groups called au-

Introduction

". . . you must admit that whatever you and your civilization
due to me—insomuch that if I had not had this dream you woul
had no existence whatever."
                                    —Julian Hawthorne, "June, 1

New technologies is a historically relative term. We are not the first
generation to wonder at the rapid and extraordinary shifts in the dimension of the world and the human relationships it contains as a result
of new forms of communication, or to be surprised by the changes
those shifts occasion in the regular pattern of our lives. If our own
experience is unique in detail, its structure is characteristically modern.
It starts with the invention of the telegraph, the first of the electrical
communications machines, as significant a break with the past as printing before it. In a historical sense, the computer is no more than an
instantaneous telegraph with a prodigious memory, and all the communications inventions in between have simply been elaborations on
the telegraph's original work.

In the long transformation that begins with the first application of
electricity to communication, the last quarter of the nineteenth century
has a special importance for students of modern media history. Five
proto–mass media of the twentieth century were invented during this
period: the telephone, phonograph, electric light, wireless, and cinema. This period is not the usual starting point for the social history
of Anglo-American electric media, which is generally assumed to begin only with the institutional birth of film and broadcasting and the
development of large audiences in the twentieth century. The present
study modestly attempts to push back those beginnings to the late nineteenth century, when Anglo-American culture was fascinated by the
communicative possibilities of the telegraph, the telephone, and the
incandescent lamp—choices that may come as a surprise to contemporary sensibilities focused on twentieth-century mass media.

diences from voiceless collectivities and to inspire new uses based on novel technological properties. When audiences become organized around these uses, the history of a new medium begins. The model used here is different. Here, the focus of communication is shifted from the instrument to the drama in which existing groups perpetually negotiate power, authority, representation, and knowledge with whatever resources are available. New media intrude on these negotiations by providing new platforms on which old groups confront one another. Old habits of transacting between groups are projected onto new technologies that alter, or seem to alter, critical social distances. New media may change the perceived effectiveness of one group's surveillance of another, the permissible familiarity of exchange, the frequency and intensity of contact, and the efficacy of customary tests for truth and deception. Old practices are then painfully revised, and group habits are reformed. New practices do not so much flow directly from technologies that inspire them as they are improvised out of old practices that no longer work in new settings. Efforts are launched to restore social equilibrium, and these efforts have significant social risks. In the end, it is less in new media practices, which come later and point toward a resolution of these conflicts (or, more likely, a temporary truce), than in the uncertainty of emerging and contested practices of communication that the struggle of groups to define and locate themselves is most easily observed.

Electrical and other media precipitated new kinds of social encounters long before their incarnation in fixed institutional form. In their institutionally inchoate manifestations, they inspired energetic efforts to keep outsiders out and insiders under the control of the proper people. Chaotic and creative experiments with new media and thought experiments with their imaginary derivatives attempted to reduce and simplify a world of expanding cultural variety to something more familiar and less threatening. That impulse fixed on one-way communication from familiar cultural, social, and geographic perimeters as a preferred strategy to two-way exchange, with its greater presumption of equality and risks of unpredictable confrontation. Classes, families, and professional communities struggled to come to terms with novel acoustic and visual devices that made possible communication in real time without real presence, so that some people were suddenly too close and others much too far away. New kinds of encounters collided with old ways of determining trust and reliability, and with old notions about the world and one's place in it: about the relation of men and women, rich and poor, black and white, European and non-European, experts and publics.

Discussions of electrical and other new forms of communication in the late nineteenth century begin from specific cultural and class assumptions about what communication ought to be like among particular groups of people. These assumptions informed the beliefs of nineteenth-century observers about what these new media were supposed to do, and legislated the boundaries of intimacy and strangeness for the close and distant worlds they presented to their audiences. How new media were expected to loosen or tighten existing social bonds also reflected what specific groups hoped for and feared from one another. Finally, concerns about how practices organized around new media would arbitrate the claims of antagonistic epistemologies contending in the public arena were rooted in group-specific beliefs about how the world could be known, and how other groups than one's own imagined it to be. Those who wrestled with these puzzles did not think in terms of the articulated mass media we know, since these inventions were still experimental and their exact shapes vague in the public and expert mind. They thought in terms of devices doing duty in familiar surroundings: the telephone, electric light, phonograph, cinema, wireless, and, always in the background, the telegraph.

This study focuses especially on two inventions on this list that have been regarded as least relevant to twentieth-century media history. The first is the electric light, which is ordinarily not thought of in connection with communication at all. The second, the telephone, has not been considered a medium of *mass* communication. Nevertheless, the telephone was the first electric medium to enter the home and unsettle customary ways of dividing the private person and family from the more public setting of the community. The electric light was the great late-nineteenth-century medium of the spectacle, dazzling its audiences with novel messages. In much social imagination, it was the premier mass medium of the future. Because the telephone and the electric light were the most technically and socially developed communications devices in the last quarter of the nineteenth century, experts and laymen found them good to think, to paraphrase Lévi-Strauss, about what media systems of the future and the societies that supported them might be like. They were also the most widely experimented with.

It is impossible to separate public discussion of innovations in communication in the late nineteenth century from public fascination with the fruits of electrical possibility generally. This is partly because "electricians" and their associates were the earliest users and closest observers of electric media. Media historians have scarcely noticed this convergence. Focused on the point of mass production, artifactual communications history has failed to recognize that electricians were as deeply involved in the field of cultural production as in the field of technical production. Technological historians also have treated electricians exclusively as technical actors, accepting mostly at face value the boosterism of their professional rhetoric. As citizens with attachments to families, communities, and social amenities as strong as any that connected them to their profession, their role was somewhat different, however. The stamp of society on them was nowhere more visible than in their uneasiness about the impact of new media on family, class, community, and gender relations. The ambivalence that so much characterizes contemporary regard for electronic media did not originate with twentieth-century radio and television, but in threats to social interaction set up by their nineteenth-century prototypes.

The temptation to derive social practice from media artifact has also supported another notion, common to media analysis, that separate media embrace distinct and self-contained codes, or spheres of interpretive activity. Concrete arenas of communication are always more complex than this. In the late nineteenth century, oral-gestural and literate codes were both projected onto electrical devices and events in the struggle to claim and label these new and important objects for social consumption. In general, literate practices were the self-consciously exclusive domain of electrical experts. To be an expert was to have knowledge based on technical texts. We can learn a great deal about how electricians and other social groups constructed the social world by observing their uses of texts, and their evaluation of others' uses as well. Groups without recourse to special textual expertise approached the electrical unknown directly, learning with their bodies what it was and what their relationship to it should be. Though deeply distrusted by experts as an instrument of naïve empiricism or folk wisdom, the body was a popular probe for making strange phenomena familiar. Even experts found it difficult to resist.

Many of the stories that constitute the evidence for this study describe real events. Others do not, but were treated by contemporaries as if they did. Still others are unselfconsciously extravagant media fantasies. This is as it should be, since fantasies and dreams are important human products that define limits for imagination. Fantasies help us determine what "consciousness" was in a particular age, what thoughts were possible, and what thoughts could not be entertained yet or any-

more. The point frequently has been made that private dreams are systematic in content and impulse. Dreams and fantasies created, exchanged, and reworked in the public forum are systematic as well.
They develop their own traditions in the conversation society has with
itself about what it is and ought to be. Such dreams are never pure
fantasy, perhaps, since their point of departure is a perceived reality.
They reflect conditions people know and live in, and real social stakes.

This exercise in communications history is not, in sum, a history
of media in the usual Laswellian sense of the set of sluices through
which societies move messages of particular types. Media are not fixed
natural objects; they have no natural edges. They are constructed com-
plexes of habits, beliefs, and procedures embedded in elaborate cultural codes of communication. The history of media is never more or
less than the history of their uses, which always lead us away from
them to the social practices and conflicts they illuminate. New media
broadly understood to include the use of new communications technology for old or new purposes, new ways of using old technologies,
and, in principle, all other possibilities for the exchange of social
meaning, are always introduced into a pattern of tension created by
the coexistence of old and new, which is far richer than any single
medium that becomes a focus of interest because it is novel. New media embody the possibility that accustomed orders are in jeopardy, since
communication is a peculiar kind of interaction that actively seeks variety. No matter how firmly custom or instrumentality may appear to
organize and contain it, it carries the seeds of its own subversion.

If new communications devices were vehicles for navigating social territory in the late nineteenth century, it is clear that some of the
maps constructed for them are fabrications we have sought to dismantle in the twentieth. It is useless to scold nineteenth-century engineers
for their failure to be twentieth-century feminists or champions of civil
rights, but it may be useful to understand how electrical experts and
their publics projected their respective social worlds onto technology
in the late nineteenth century, and what justifications and fears moti-
vated them in this. It is also important to notice that communications
technologies that prepared the way for twentieth-century media were
built to uphold a scheme of social stratification that has attracted sustained contemporary challenge. This, as much as anything else, is a
measure of how we have changed.

1

Inventing the Expert

Technological Literacy as Social Currency

> The poem is a standardised one, in that it is passed on in a particular context; by selected people and in a special style, people are encouraged to listen to and then recite the myth and a premium or reward is given to those who can do this well.
>
> —Jack Goody, *The Domestication of the Savage Mind*

> "Any opinion mankind has held that has not been through the crucible of science is probably wrong."
>
> —A[mos] E. Dolbear, "The Science Problems of the Twentieth Century," *Popular Science Monthly*, 1905

Electrical professionals were the ambitious catalysts of an industrial shift from steam to electricity taking place in the United States and Western Europe at the end of the nineteenth century. According to Thomas P. Hughes, Alfred Chandler, and others, that shift was made possible by key inventions in power, transportation, and communication, and by managerial innovations based on them that helped rescale traditional systems of production and distribution.[1] The retooling of American industry fostered a new class of managers of machines and techniques; prominent among them were electrical professionals. The transformation in which these professionals participated was no class revolution, as David Noble has pointed out.[2] Their job was to engineer, promote, improve, maintain, and repair the emerging technical infrastructure in the image of an existing distribution of power. Their ranks included scientists, whose attention was directed to increasingly esoteric phenomena requiring ever more specialized intellectual tools and formal training, electrical engineers, and other "elec-

9

tricians" forging their own new identity from an older one of practical tinkerer and craft worker. Servingmaid to both groups were cadres of operatives from machine tenders to telegraph operators, striving to attach themselves as firmly as possible to this new and highly visible priesthood.

Electrical experts before 1900 were acutely conscious of their lack of status in American society relative to other professional groups.[3] The American Institute for Electrical Engineers (AIEE), founded early in 1884, was the last of the major engineering societies to be organized in the nineteenth century.[4] Professional societies had already been organized by civil engineers in 1852, mining engineers in 1871, and mechanical engineers in 1880. The prestige of other groups in the engineering fraternity, especially civil and mechanical engineers, came less from membership in professional societies, however, than from other circumstances. Their practitioners hailed from the upper and middle strata of society, were often products of classical education, and had developed distinctive professional cultures of their own well before the formation of their national organizations. This gave them an established and even aristocratic niche in society.[5]

None of this was true for electrical engineering, which had emerged only in the decade before the founding of the AIEE, and which by the time of its organization had achieved no clear consensus about the meaning of the term *electrical engineer*. The broader title *electrician* was equally vague.[6] It appeared as a distinct census category for the first time in 1860, but despite a flourishing telegraph industry, only 12 practitioners were reported. Not until 1900 were electricians mentioned separately again, when 50,717 workers were so classified.[7]

Before 1900, as Robert Rosenberg has written, the electrical work force comprised a motley crew from machine tenders to motor designers and from physicists to telegraph operators, all sharing in some fashion the title *electrician*.[8] Anyone interested in electricity might claim it, and many did. "It is doubtful whether any man present over thirty years old selected any application of electricity, with the exception of the telegraph, as a means of livelihood in the sense that a youth would select a trade or . . . professional avocation," one of those professionals reminded his colleagues at the first annual meeting of the Electric Club in New York in 1887.[9] His exception for telegraphy was not much of an exception, since telegraph operators enjoyed scant occupational prestige compared with other electrical professionals.

A number of trade and technical journals were witness to the occupational and status anxieties of electricians. The first general weekly electrical paper for professionals in the United States was *Electrical Review,* founded in 1883. *Electrical World,* perhaps the major electrical industry journal in the United States in the late nineteenth century, claimed the largest circulation and boasted more than seventeen thousand readers by 1895. These and other journals like *American Electrician, Electrical Engineering, Western Electrician,* and, to a lesser degree, popular science journals like *Scientific American* kept readers abreast of the latest in electrical innovations and scientific findings bearing on their craft, promoted and recorded professional meetings and activities, and commented on affairs of industry and politics that affected the electrical profession. In contrast to the loftier AIEE *Transactions,* these journals addressed not only academic and practicing scientists and engineers, but also foremen, superintendents, designers, managers, entrepreneurs, and other workers in the field of commercial electrical application. Without exception, these journals subscribed to the argument that electrical experts were entitled to greater social position and respect, a quest officially framed as the pursuit of proper standards and career experiences for training future electrical workers.

Scattered throughout the technical reports and documents that constituted the primary focus of this literature was a secondary content of social news, editorial comments, and short anecdotal articles that provided a less earnestly self-conscious arena of discussion. Its ostensive subject matter was the movement of an expanding and varied culture of electricity through the larger society. It included excerpts from the lay press, material quoted incestuously back and forth from other journals—a widely acknowledged and generally accepted practice—and tales attributed to every imaginable source. The casual tone and location of this material, at the interstices of the strait-laced technical and professional documents which announced that electricians were busily engaged in their calling, made it ideal for expressions of the concerns closest to their hearts.

The industry in which these workers labored, and to which their concerns were directed, was significant and growing. At the beginning of 1890 one journal estimated that $600 million had been invested in the electrical industry of the United States, 250,000 people depended on it for their livelihood, one million miles of telegraph wire had been strung ("enough to circle the globe 40 times," crowed one expansionist metaphor), and 1,055,500 telephone messages from 300,000 instruments were daily buzzing over 170,000 miles.[10]

Electrical Textuality

Brian Stock has given us the term *textual communities* to describe groups
that rally around authoritative texts and their designated interpreters.[11]
Stock's work addresses certain realignments of medieval discourse, in
particular what he regards as an original divergence between popular
and high culture. His notion provides a useful starting point for considering other textual communities, their spokespersons and interpreters, and their relationships to less lettered communities. In the late
nineteenth century, aspiring electricians placed scientific textuality and
certified interpreters of scientific texts at the center of their claim to
public authority, and attempted to persuade those less technically lettered of the validity of that strategy.

The notion of scientific textuality appeared over and over in discussions of professional standards. The editors of the *Electrical Review*
praised the young American Institute of Electrical Engineers for the
"large number of valuable papers touching upon almost every branch
of the electrical industries," and expressed concern that the level of
discussion at meetings of that and other societies, including the New
York Electric Club and the older New York Electrical Society, was
rarely up to the level of the papers themselves. They urged technical
societies to bear in mind "that the proceedings are read and studied by
electricians the world over."[12] Documentary procedures were so central to electrical engineering practice and research that it is not inac-
curate to use the term *technological literacy* to describe a range of
professional competencies that at their core valued skill in interpreting
technical documents. Electrical engineers and researchers fully intended that these literate skills and the theoretical knowledge they embodied replace the skills of the tinkerer and craft mechanic, skills governed by an authority of the body that arrives at truth from the direct
experience of the senses.

Broadly speaking, four communities accepted the expert authority
of electricians and their texts in the late nineteenth century, or at least
were addressed by electricians as if they did. Together these communities were organized around a presumptively shared, but distinctively practiced, epistemology of texts and interpretive procedures that
were sanctioned by certified authorities arranged in roughly concentric
circles of expertise. First was the select readership of theoretical and
entrepreneurial electricians addressed by every kind of professional and
technical literature. Professional societies were important to this tex-

tual community as well, since most of their meetings were referenced to texts around which the mutual interests of their members revolved. A second textual community collected around the literature of popular science that aped conventions of expert presentation, and sometimes the mantle of professional and scientific authority as well. This was the group, explained the authoritative British *Electrician,*

> whose earnest efforts give a far greater publicity to our notes and to many of our articles than we ever contemplated. These people are more accustomed to wield the paste and scissors than the pen, and we presume it is due to their lack of familiarity with the latter auxiliary that they so seldom mention the title of the paper to which they are indebted for their matter.[13]

Suspiciously monitored by the professional press for sensational tendencies, this community aimed primarily at a popular audience of enthusiasts. The circle of interpreters it accepted as legitimate was larger, looser, and less differentiated than in the more strictly accountable professional press.

A third community was constituted in the flow of information, characteristically in one direction, from electrical experts as accredited interpreters directly to lay audiences, generally of the middle class. It made itself heard in the oral channels of lecture and lyceum, and in articles written for middle-class literary journals like *Fortnightly Review.* This was the audience idealized in a description of a standing-room-only crowd at the Royal Institution on the occasion of a lecture on wireless telegraphy by Guglielmo Marconi:

> As usual, the assembly was a mixed one, from our neighbour who regretted he had not had time to read up the subject in the "Encyclopaedia Britannica" beforehand, to the scientist who came with the hope of hearing the announcement of a new discovery. The audience also included a large proportion of the fairer sex, a number of whom were old *habituées.*[14]

These exchanges were disseminated to a still larger audience by the popular press, which often reported on these occasions, but rarely in a manner satisfactory to expert eyes. To the dismay of electrically literate elites, the popular press embraced colorful charlatans as enthusiastically as it did certified experts. This popular press and its electrically unlettered audience constituted a fourth textual community. From time to time the professional electrical press offered the gatekeepers of the popular press suggestions for improvement. "Although we have never been enthusiastic advocates of science for the multitude," wrote the *Electrician* in 1882, "we would certainly make an exception in favour of newspaper editors. In the interests of the public, for whom the journalist professes to live, he might, one would think, include a smattering of science in his professional training."[15]

Of special interest is Stock's account of the challenge to religious orthodoxy mounted by heretical and reforming communities that took the principle of textual authority to heart, but applied its logic in new and unanticipated ways. Debates over competing interpretations of sacred text brought the communities sponsoring them into conflict. Their disagreements were rarely about the priority of textual authority, or even about broad principles of legitimate interpretation. Their differences concerned substantive points of interpretation and the doctrinal implications of these differences, not least among them disagreement about the valid sources of religious authority in this world.[16] If the community of electrical professionals had less at stake than the medieval church, it too was challenged by the very groups it hoped to convince of its unassailable textual authority, and this for the simple reason that it had made electricity too fascinating a topic for popular culture to leave alone.

A recurring theme in the study of literacies past and present is how skills and techniques for performing particular literate practices are transferred from communities of adepts to less skilled communities. What is not so easily transferred is the specific cultural setting and world view that gives significance to these practices from the point of view of the bequeathers.[17] This is part of the historical irony by which medieval religious elites were beset by the very groups they had intended to control. Borrowing elite rules of interpretation, these less powerful groups constructed a textual exegetics shaped to their specific needs and experiences of the world. Wherever their interpretations were resisted by established textual communities, believers in textual authority took on, often fiercely, those who had taught them the importance of the principle. Confronting a similar if less intense challenge, late-nineteenth-century electricians stood guard over popular efforts to interpret electrical phenomena in ways that seemed to undermine the credibility of scientific experts. Though generally convinced of both the magic efficacy of electricity and the authority of the magicians who manipulated it, popular interpreters drew their own conclusions when it suited them.

But this is the limit of the analogy. Where Stock's concern is with a world in which reverence for textual authority inspired those who were disbarred from membership in elite textual communities to invent a *popular* textual culture, our concern is with the effort of electrical professionals to invent themselves as an *elite* in the late nineteenth century. To this end, much of the literature of electrical mission was occupied with sorting and labeling insiders and outsiders in electrical culture. Technological literacy, in the sense defined here, was critical evidence for such distinctions. The proper naming of persons, gadgets, and concepts in their electrical contexts and relations was among the most important performative indicators of technological literacy, even though contemporaries coined no distinct term for this skill. What it meant to possess the skill of electrical naming and understanding was worked out in thousands of examples in the literature, all of which explored codes of meaning attached to electricity in society. Absent this contemporary effort to take the *social* measure of technological literacy, specific technical skills and performance criteria could have no real existence or application.

Occasionally, those outside the boundaries of textual demarcation fashioned by experts refused to defer to those limits or recognize the social and professional privileges attached to them. When this happened, deception of the less by the more literate was considered an acceptable and even necessary option to keep these boundaries secure. The professional literature exhibited scant interest in whatever ethical questions might be involved in deceptive manipulations to achieve power over the less expertly informed. Most of the time, such maneuvers were not even explicitly defended, since knowing when and how to execute them was a marker of group solidarity, the more so the more restricted and exclusive the level of electrical literacy.

Insiders and Outsiders

Much of the electrical literature described above and a significant portion of the technical literature it supplemented explored social relations between electrical insiders and outsiders around textual concerns. Electricians were wont to indulge a powerful impulse to identify aliens and enemies, those suspect in electrical culture and perhaps dangerous to it, in terms of their textual competence. Outsiders were defined as those who were uneasy and unfamiliar with technical procedures and attitudes, especially literate ones. By a supplemental logic of explicit social control, any additional marginality of race, class, gender, or lifestyle was taken as confirming alien status. The effort to identify outsiders by textual cues naturally raised the reverse issue, namely, who had legitimate claim to the title *electrical expert,* and by what literate deeds they could be recognized and certified in the expert arena and in society at large. The literature of electrical mission also occupied itself with the problem of what legitimacy to confer upon an admiring public's efforts to interpret the world of electrical science and engineering, especially when the conclusions it reached ranged far afield of textually disciplined expert notions, and especially when experts' own goals were to harness public adulation to improve their own social and professional standing while keeping public admirers at arm's length. One official boundary at which electrical insiders and outsiders met was negotiated in a currency of promises given by insiders to outsiders, that is, by experts to publics, and equally in expectations held by laymen concerning their right to share in an electric prosperity made possible by public recognition and indulgence of expert ingenuity. Expert and popular literature alike monitored the rhetoric of reciprocity, watchful for any breach in the vague but binding bargain between experts and their publics in behalf of electrical progress. Experts, for their part, frequently took their erratic publics to task, as often for believing too little as for believing too much.

Electrical experts attended to several gross indices of technological literacy. An ad for an "Experienced Electrical Engineer" in one journal sought an aspirant "well up in Electro-Mechanics, good at experimenting and technical reports." Documentary skill was thus cited as a fundamental professional qualification, and being "well up" on electromechanics implied an ability to follow the latest technical literature. "Only one person out of every two thousand in this country reads the electrical journals," *Electrical World* estimated in 1889, surmising as to what the size of the community of electrical literates might be.[18] The *Electrician* portrayed a fictional proprietor praising his newly hired engineer for both his electromechanical skill and the command of literate procedure that flowed from his specialized textual knowledge.

> "How does your electrical engineer go on?"
> "Oh, very well, we never know what a break down is since he came, and if we want to make any alterations or to put up any new apparatus . . . he brings me the order to sign, or gives his estimate, and that is all I know till I see the thing working."[19]

Claims to expertise on the basis of textual credentials could be challenged if the claimant were clearly a social outsider, since textual cues were expected to signify appropriate social circumstances. "A dirty-looking young man once called upon us," the London *Telegraphist* wrote, "handing a well-thumbed type-printed card, bearing the unwashed one's name, followed by the word *Electrician*."[20] The young man had presented textual evidence worthy of consideration, but nothing else was consistent. His appearance made his claim suspect, and his name did not connect him to a network of familiar insiders. The verdict of these signs was confirmed in the final test, which revealed the young man's conversation to be technically improficient. He was an electrician by textual pretense alone, utterly lacking the extratextual finish assumed to accompany authentic technological literacy.

Even laymen were expected to possess some literate skills for coping with electrical technology. Those who were socially positioned to know this assumed inventive poses if their skills were not up to par. A "quite respectable-looking young woman" asked the receiving operator to write down her telegraph message for her, since she could not do it herself with her gloves on. Her ruse implied a minimum standard of literacy expected of an enlightened citizenry for coexisting with the practical aspects of electricity, and clearly associated with other visible signs of class. This story also portrayed telegraph operators as a highly literate lot, admirably sensitive to these class cues by virtue of their occupation, and possessors of an admirable humanity that provided a showcase for technical prowess:

> It is quite a common thing for people, both men and women, to ask us to do their writing for them. I guess anyone would be astonished to find out how many people there are who are hardly able to spell their own name, much less write a legible letter or telegraphic message. These are principally English people of the working classes, who have only been in this country a short time. Nearly all born Americans can write. They tell me that in England the laboring people are very seldom able to read and write, especially in the mining and manufacturing districts. . . . They will pretend . . . they have sprained their wrists, or have their gloves on, or can't write with our pens, and we have to look serious, while all the time we see through their dodges perfectly.[21]

Stigmatizing the Unempowered: Rural, Female, Nonwhite

The professed goal of authoritative discourse in electrical journals and at conventions was to debate technical problems and to discuss whatever social and professional concerns might bear on them. Electricians did not hesitate, however, to extend their concerns beyond the boundaries of professional culture, though they did not consider their own preserve equally permeable to opinion from without. To electricians, other social groups were faintly contemptible, definitely so if their members ventured into unfamiliar expert territory.

Criteria for distinguishing electrical insiders and outsiders were clearest in jokes of internal cohesion that provided light features and filler in the electrical press. They poked fun at how outsiders attempted to navigate codes and procedures electrical insiders took for granted. The usual targets of this humor were black, foreign, rural, or female, despised groups in the system of caste that experts shared with the larger society. Persons of rank and privilege were capable of earning the hostility of electricians, but never appeared quite so ridiculous as those who provided a readier target for social scorn. An official of the Edison General Electric Company recalled that he and Thomas Edison had once called on one of New York's "biggest" millionaires to discuss installing electric lights in the millionaire's mansion. During the conversation, the millionaire asked whether Edison could install an electric motor to run the steam engine that operated his passenger elevator.[22] This was a joke, but a mild one. Its narrator was only bemused by what "the outside world knows about electrical matters"; comments about less exalted groups were more likely to elicit complaints about the futility of expecting marginal groups to understand and appreciate what electricity could offer. In their efforts to reorganize a social hierarchy with no definitely settled place for them, experts sometimes measured themselves against those whose power they expected to decline in a world of new forms and correspondingly new structures of influence. The trade journal *Lightning* pilloried diplomatic verbosity, a traditional signifier of aristocratic social class and high political authority, as incongruous in the telegraphic domain, into which diplomacy had begun to pass from the more dignified arena of oral and written exchange:

> What a magnificent thing it would be for the Post Office if everyone telegraphed at the same length as certain Emperors and Princes. "William" contrived to get 112 words into a simple message to Bismarck to the effect: "Only just heard of your illness. Come and put up with me"; and Bismarck broke his record by telegraphing in 206 words the reply: "Thanks. Sorry it cannot be managed."[23]

Still, jokes in the electrical press were aimed mostly at those with little social power, occupying either the conditions of misery that electrical progress was supposed to alleviate or positions that would have to move aside to make room for electrical success. In asides and anecdotes, electrical experts thus defined themselves as much by the groups from which they chose to disassociate themselves as by those with whom they sought alliance. The *Albuquerque Journal* narrated the story of Royal Wilson, a black man elevated, by the sudden illness of the headwaiter in the hotel where he worked, to his boss's post. When it was time to extinguish the electric lights in the dining room, Wilson, a man cast loose from his social moorings, found himself in a state of "painful uncertainty." He decided, explained the *Journal* with malicious irony, "that the simplest way out of a difficulty is always the best."[24] Leaning precariously from a chair perched on a table, he blew "until his eyes bulged out and the sweat trickled in rivulets from his features." This image was a familiar racial stereotype, and these were the desperate gestures of one to whom a technology based on something besides muscle power was an impenetrable mystery.

Not knowing how to turn off the lights was a familiar comic theme. A cartoon in an illustrated paper showed Uncle Hayseed in a New York hotel inverting his large, rude boot over the lamp after many futile attempts to blow it out.[25] In another story, a puzzled rancher at a Seattle hotel finally succeeded in uncoiling the wire from which the lamp in his room hung, so that he could stuff it into a bureau drawer to extinguish it.[26] Humor at the expense of powerless groups established a social floor above which electricians felt comfortably smug. The professional journal-reading community could bask in the social assurance of their own society pages, since their journals were read by a small, mutually acquainted community.

Other stories contrasted rural credulity with urban sophistication, and satirized practitioners of mechanical technology who seemed unable to accommodate electricity:

> The telephone is a puzzling mystery to the rural mind that tackles it for the first time. For instance, a countryman approached a telephone man in Boston the other day with the following interrogation: "Now, mister, what makes the thing work? Thar's yer wire and thar's that 'er trumpet and all that, but ain't thar suthin' aside o' that? Whar's the steam, the *push* to the thing? What makes the talk go 'lang so? *What greases the derned thing?*"[27]

The joke is on the bumpkin who clings to his anachronistic mechanical model of technology in a world where reasonable people know better. His status as an outsider is manifest in this error. To underline that status unmistakably, his ungrammatical dialect appears in pointed contrast to an elite facility with genteel language and expression, and it is implied that electrical experts, as readers of the story, belong to this more desirable group.

The outsider as stock rural character appeared in the *Sacramento Record-Union* as a "raw California granger" in a story about the social mischief of technological ignorance. The story is presented by an omniscient narrator who occupies a logically impossible vantage point for observing the mutual frustration of granger and expert without either's knowing the full set of story events. The story is a moral fable of social relations borrowing the dramatic force of a putatively factual account. A reluctant granger found it necessary to use the telephone. He approached it "timidly," eyed it "cautiously," and, taking a pencil, began to write on a piece of paper.

> He then rolled up the paper and tried to push it in the aperture in the transmitter. Failing in his attempt with his finger, he took his lead pencil and jammed it in, destroying the vibrating plate. With an air of satisfaction he took his seat and awaited a reply. After about ten minutes he became discouraged, and thinking he perhaps had not sent the message on the right line, he wrote another and jammed it into the hand telephone, and to make sure work, rammed it home as he would a ball in a rifle.[28]

The puzzled granger departed after another half-hour wait, and a secretary entered the room. He discovered the telephone "stuffed full of manuscript and ruined." When the instrument was dismantled and all messages had been removed, they were all found to read: "Bakker and Hammeltonn—Send me to the Pavillion a six inch long munkey rench. Yurs Trully J. E."

The granger signifies an economic order attached to the land and wedded to inelegantly mechanical procedures unsuited to the complexities of electricity. The granger is doubly illiterate, and this makes him dangerously destructive in a technically sophisticated world. Not only are his actions premised on an incorrect analogy between written literacy and the telephone; he is not even proficient in the written literacy on which his actions are modeled. He is also a threat to property, though the electrical order is ultimately victorious, since he must pay for the damage he causes. The proprietor of the telephone fences it off with a "Beware of the Dog" sign to deceive functional literates who, like the granger, lack the critical capacity, associated with more sophisticated literate skill, to question what they read. By an unspoken principle that informs all this literature, the technologically marginal are deemed deserving of deception at the hands of those with greater skill.

Along with textual competence, other gross indicators of technological literacy included skill in operating electrical machinery and, always, sensitivity to the social conditions and constraints surrounding the exercise of those skills. Unhesitating appreciation of the virtues of new electrical technologies and the experts who oversaw them completed the list. In the realm of electric communication, this last condition implied an absolute belief in its uniqueness, and the refusal to entertain any notion that electric communication merely extended or speeded up oral and written communication, or was an equivalent substitute. By its very nature, in other words, it was not subject to existing social rules. It was truly new, and rules for using it owed nothing to the past, but only to engineers bent on creating the future. It was a short step from perceptions of electrical communication as a phenomenon outside the realm of personal or cultural values to the conclusion that expert-prescribed instructions for its use were not the mutable product of human custom, but given in nature itself.

To agree with these facts as electricians understood them was to embrace a model for prosecuting electrical communication with brevity and efficiency. Obedience to it distinguished those whose "correct" perceptions encompassed a larger, more sophisticated world of technology from those whose imaginations played on smaller, less impressive stages. Typical was the story of a baker's assistant whose wife was gravely ill, and who seized on the telephone as just the thing to persuade his sister-in-law to come home at once. He rushed to his former employer's establishment and asked to use the instrument there. This detail emphasized the main point, the social distance between the technologically initiated and uninitiated. Permission granted, the butt of electrical amusement stepped up to the telephone. "Then without ringing up the central station and getting connection, without taking down the ear tube, he just hallooed into the hole: 'Kitty, come home! Mary's sick!' and vanished before anybody could stop him."[29] It developed that Kitty could not be reached by telephone where she worked, but such had been her brother-in-law's faith in the telephone that he thought "all he had to do was to speak into the instrument and it would carry the message anywhere he desired."

Electricians were amused at the miraculous powers vested in devices for electrical communication by the technologically naïve. These powers displayed the features of the oral and written models they were based on, few of the unique capacities of electrical communication, and additional magical capabilities that to experts were inconceivable for *any* mode of communication. A popular misconception was that telegraph and telephone messages were written down and physically transported over the wire. In the earliest days of telegraphy, "even fairly educated people believed that the paper passed along inside the wire," reminisced a British railwayman in 1890.[30] Now, he implied, only the most socially marginal could make this error.

Some enthusiasts imagined that electrical communication was mysteriously enhanced oral discourse in which speakers and listeners were seen as well as heard, just as if their conversation were face-to-face. In one story, an office boy in a business house in Aberdeen, a "raw country youth" speaking the patois of humble station, was minding the telephone in his master's absence.

> When first called upon to answer the bell, in reply to the usual query, "Are you there?" he nodded assent. Again the question came, and still again, and each time the boy gave an answering nod. When the question came for the fourth time, however, the boy, losing his temper, roared through the telephone:
>   "Man, a' ye blin'? I've been noddin' me heid aff for t' last hauf 'oor!"[31]

Featured in many stories was the frustration of the technologically unempowered, expressed as anger, fright, or other loss of personal control. These displays contrasted with the cool bearing of the professional, whose perfect awareness was accompanied by an equally flawless emotional control that suggested social and moral superiority. Uncontrolled emotion was displayed by men who were victims of their own technological ignorance, who had somehow shirked their responsibility to be technologically informed.

The Special Case of Women

Women's ignorance, on the other hand, was ignorance even of the extent of their electrical incapacity.

> A gentleman, talking with a young lady, admitted that he had failed to keep abreast of the scientific progress of the age. "For instance," said he, "I don't understand how the incandescent light, now so extensively used, is procured." "Oh, it is very simple," said the lady, with the air of one who knows it all. "You just turn a button over the lamp, and the lights appear at once."[32]

Technical ignorance as a form of worldly ignorance was a virtue of "good" women, as they invariably were in the professional literature, where encounters with "bad" women were not discussed. Unlike men, women in the stories related by professional journals rarely learned from their mistakes in using technology, or corrected their misconceptions. They were sheltered from all such practical demands by an old and sturdy code of chivalry that required the protection of their ignorance by men. Beneath this habit of indulgence was the more important and even insistent point that women's use of men's technology would come to no good end. In keeping with the general portrait of women as impotent, even their most exasperating errors usually had little more consequence than inconvenience to themselves, of which they were varyingly aware, and some slightly larger measure of frustration and inconvenience for their male protectors.

In the picture painted by electrical journals, the model of electric communication that came naturally to women and led them astray was the loquacious oral sociability of their everyday lives. Talkative women and their frivolous electrical conversations about inconsequential personal subjects were contrasted with the efficient, task-oriented, worldly talk of business and professional men. A hypothetical telephone conversation between two women in the *Electrical Review* of 1887 demonstrated the incomprehensibility of the telephone to a feminine construction of the world. The conversation began this way:

> Mrs. Wary (at the telephone)—"Hello, hello, Exchange." After waiting some time without a reply, Mrs. Wary, in more vigorous tones, pipes out "hello." Still no reply, whereupon Mrs. Wary softly murmurs so that the telephone will not hear her, "Well, I declare, if I don't believe I forgot to ring. How stupid." Which was a fact. Mrs. Wary then rings with a vigor and persistence without doubt intended to make up for her previous omissions, and is answered by the exchange.
>
> "Connect me with number—number" (in an aside) "bless me but I've forgotten the number," (she so informs the exchange, but is finally put in communication with her friend, Mrs. Prim, when the following conversation ensues):
>
> Mrs. Prim—"Is that you, Mrs. Wary?"
>
> Mrs. Wary—"Why, of course, it is. How did you happen to call me up, I was just going to call you up. Isn't it nice."[33]

The women discuss the good looks of several local pastors and gossip about fashion and dressmaking. To experts their conversation is trivial and uninformative, and could be as easily managed face-to-face. At

the end of the conversation their failure to understand the urgent and
serious nature of telephone talk is especially clear.

> Mrs. Prim—". . . But what a nice talk we've had. It's a wonder
> that the horrid girl at the exchange has not shut us off before this time."
> Mrs. Wary—"So it is. I've forgotten now what I called you up
> for, but I guess it's of no consequence, so good-bye."

Women appeared as the parasitic consumers of men's labor in
most stories of their electrical ignorance. Many of these stories turned
on wives and girlfriends instructed to send telegrams or make telephone calls to reassure those charged with their care of their safe arrival at distant destinations. Predictably, these women failed to understand electrical messages the way their male protectors did, as scarce
and expensive commodities. To women, electrical talk was a delight-
fully extravagant extension of face-to-face intimacy, almost a free good.
Men found themselves caught by their obligation to a traditional code
in which women were not supposed to understand the stern masculine
world of electrical knowledge, while men were supposed to live by its
rules. Men were forced either to choose the displeasure of the women
they loved or to pay profligate sums incurred by wives, girlfriends,
and sisters for lengthy telegrams and phone calls. Chivalry bade them
choose the second alternative, and this financial sacrifice, characteristic of modern knighthood, was appreciated least of all by the women
for whom it was made.

In contrast to men, women valued conversation that was redun-
dant, frivolous, playful, and abundant. Such excess bespoke an affec-
tionate devotion to their partners, manifested in a generous willingness
to communicate. In return, they wanted their male partners to speak
to them the same way. For women, instrumental information about the
world outside the personal relationship that was the real subject of any
electrical conversation was irrelevant. Women regarded the brief, efficient transmissions prized by men as an evasion of the relationship
that they assumed it was the point of any communicative exchange to
cement.

Men, by contrast, wanted control of all communication conducted
through the technology that belonged to them. Rules of expertise that
invested the knowledgeable with power over the less knowledgeable
transformed stories of women's electrical ineptitude into homilies that
justified men's control of women's communication. *Chambers's Journal* published a story from the "infancy" of the telegraph about one

elderly lady's conviction that telegraphy should follow the rules of propriety familiar to her from a lifetime of nontelegraphic communication. The telegraphist at the counter of London Central Station, "to whom it really occurred," received from this lady a sealed, addressed envelope containing the message she wished to send. She was indignant when the clerk opened the envelope, even when he explained that he could not send the message without seeing it. " 'Then,' replied the female, in evident ire, 'do you suppose I'm going to let all you fellows read my private affairs? I won't send it at all;' and therewith she bounced out of the office in high dudgeon."[34]

From a male perspective, the usual puzzles of communication between the sexes were exacerbated by technological codes that bound men but that women did not respect. Put another way, male control of female communication was justified by women's ignorance, and should have guaranteed it as well. But women often frustrated it anyway. Annie Bifkins Blank, newly wedded and visiting her mother outside Philadelphia, composed and sent her first telegram to her husband, ten dollars collect:

> Frog Center, Pa., 2 p.m.—George Washington Blank, 43 Blank Street, Philadelphia—My Dear George: I have just arrived safely without any accident at all; not the slightest. The train slowed up at Jinks crossing and whistled, but I don't think anything serious was the matter. It made my heart jump to think how you would feel if anything had been the matter, you know, but there wasn't, not a thing, so far as I could find out. I got to thinking of you and might have been carried past my station if Cousin Will, the one you used to be so jealous about, you know, hadn't been on the train. He is visiting at mother's, and is handsomer than ever. He says he hates you, but of course, that's only fun, you know. I forgot to say that my trunk came through all right. It was no trouble at all. Cousin Will took my check and arranged to have it (the trunk, you know) hauled up to the house. It will have to be taken around by the mill because the other road is blocked up, you know; but, you know, that will only take a few minutes longer than by the other road—the one that is blocked up, I mean. Well, I must close this dispatch, because telegrams have to be short, you know.
>
> Your loving wife,
>
> Annie Bifkins Blank[35]

A similar story in the *New Orleans Times-Democrat* chronicled a broken engagement that resulted from a telephonic misunderstanding. It was told, as most of these stories were, from the masculine point of view:

"I was in Atlanta a few weeks ago and called up my fiancee in Macon to let her know when to expect me. The service costs 50 cents for three minutes, and I calculated I could deliver my message in about 14 seconds. But after I gave the dear girl the date she insisted on holding me while she told about a lawn fête that some of the young people were getting up for the next day. I wriggled and writhed, and after she had imparted $2.50 worth of details I broke in and told her that somebody else wanted to use the 'phone. 'O no, they don't,' she replied, 'the operator here says you may have it as long as you wish,' and on flowed the legend of the lawn. She told me how all the girls were going to be dressed, what they had cooked for lunch, and how Annie Jones had refused to go with Billy Smith, because it was rumored that Billy played cards on Sunday. I groaned. I had been stuck for about $7, and time was flying at the rate of 16 2/3 cents a minute. 'What's the matter?' she asked anxiously: 'you don't seem interested.' 'Yes, I am,' I said, with perfect truth: 'I am weighing every syllable.' 'Then repeat what I have been saying,' she ordered; 'go all over it and don't miss a word.' That was too much. I yelled: 'Ring off!' and banged the receiver on the hook. Next day I got a package from Macon, returning the engagement solitaire. There was a sarcastic little note in which she said she thought my suggestion about the ring was excellent and had acted upon it at once. Plague take long-distance 'phones! I never want to see one again in my life."[36]

If women of fallen reputation did not exist in the electrical press, women of uncertain reputation did. Not by accident, most of them held jobs in which they operated new technology. Women were most acceptable in the labor force as austere heroines in the pioneer mold, or as devoted servants of indulgent male overseers. Otherwise, they appeared as intruders of dubious ability and fragile reputation. Either they were obedient and servile, no threat to the male world in which they moved, or they skirted the very edge of sexual propriety, a condition that released the men around them from responsibility for their welfare.

An article that instructed readers about how to recognize women telegraph operators out in the Wild West, the symbolic boundary of civilization, where the pressures of savagery against the social virtues represented by women were strongest, typified stories of women workers with selfless and saintly characters:

Far out on the western plains, wherever there is a road station, almost invariably the traveler sees a pretty lace or muslin curtain at the window, a bird cage hanging up aloft and some flowering plants on the narrow sill, or a vine trained up over the red door . . . and if he looks out as the train stops he will be nearly sure to see a bright, neatly dressed, white-aproned young woman come to the door and stand gazing out at the train and watching the passengers with a half-pleased, half-sorry air. This is the local telegraph operator, who has taken up her lonely life out here on the alkali desert amid the sage brush, and whose only glimpse of the world she has left behind her is this brief acquaintance with the trains which pass and repass two or three times during the day. These are true types . . . of our brave American girl.[37]

The woman who, nunlike, renounced the world or chose to remain isolated in her profession distanced herself from ordinary talkative women, and also did not interfere with men.

Equally virtuous was the woman who joined the electrical work force on account of reversed circumstances, who had something better in mind for herself but was the victim of a fate beyond her control, a situation ripe for rescue by men. A common theme in popular magazine fiction was the lone woman forced by circumstances, met bravely and with cheerful pluck, to make her way as a telephone or telegraph operator. At this labor she captured the heart of a good man who wooed her from that unsheltered and risky occupation to become his wife. Mention was often made of her aspirations to a more dignified station, though she seemed powerless to achieve it herself. "But surely," a Western Union manager in an 1897 short story advised a young woman who had applied for work to support herself and her widowed mother, "with your accomplishments you do not need to be a telegraphist." His applicant, a lady of the better class, replied, "My accomplishments, although expensive to buy, are not very saleable on the market."[38]

Women entered the technical world at the sufferance of men. Over and over it was made clear that they were not the help they should have been. A characteristic anecdote in the *Somerville* (Massachusetts) *Journal* concerned an imaginary conversation between Mr. and Mrs. Brown on the subject of telephone operators, the most visible female workers in the electrical industry. Why, asked Mrs. Brown, predictably the less well informed of the two, were telephone operators usually women? Mr. Brown answered:

> The managers of the telephone companies were aware that no class of employees works so faithfully as those who were in love with their labor, and they knew that ladies would be fond of the work in telephone offices.
>
> "What is the work in a telephone office?" Mrs. Brown inquired further.

"Talking," answered Mr. Brown, and the conversation came to an end.[39]

According to male testimony, women workers could not cast off the orality to which they were inclined and which made them unfit for responsible work in serious environments, though their failings were tolerated with more or less good humor by the men around them. "With a telephone and a wife a man ought to hear all that's going on," joked the *Danbury News* in England.[40]

> "Telephone girls in Chicago look black over an order to dress in uniforms of that sable color," said the Judge.
>
> "No wonder they object to black. Yeller would be more appropriate for a telephone girl's uniform," replied the Major.[41]

The exchange room of the Hudson River Telephone Company was where, the *Albany Journal* exclaimed, "15 girls chew gum and chatter all day long. What noise they make!"[42] The oral behavior of these women was the only topic of note, despite their manifest skills as exchange operators performing a range of social and mechanical tasks. Chief among them was speaking to subscribers, accusations of frivolous speech to the contrary. These workers seemed to be doing what women did best and what, judging from the way they were presented, was the only thing they could do in any case—talk. Such stories confined women's skills to an oral arena that at no point encroached on the male prerogative of technological literacy.

The power of the female telegraph operator was also carefully circumscribed. "She will sometimes have about her a number of subordinates of the opposite sex in the form of callow youths and messenger boys," explained the *New York World*, "over whom she queens it with a right royal will and an air of authority that is charming to behold." So long as it was charming. The *World* could indulge the female operator in her command of males who were not yet men, but drew the line at exhibitions of genuine power. "Generally these young women are very pleasant and obliging; only occasionally will one come across a terror, whose very look will freeze him to the marrow."[43]

A contemporary portrait of the telephone girl described her as "pretty—of course she is—she dresses with nice taste." On account of her lovely smile, she did not deserve the wrath of the "old fossil" she had inadvertently connected to an undertaker when he asked to speak to someone at the bank. This story, and many like it, cloaked the verdict that the telephone girl did her job badly in compliments to her femininity. And why not, since her job skills were less important

than the persuasiveness of her feminine charm. Unable to be taken
seriously for her technical skills or her "curious" political comments
(which were not, it seemed, her own conclusions, but gleanings of
overheard conversations), what she did know derived as usual from
her special oral skills:

> She can tell you if she wants to on what night last week young Smith's
> baby was taken sick with the colic, and how the worthy *pater* could not
> be found, but was finally discovered with a congenial party indulging
> in the fascinating game of draw-poker. But she won't tell you this if
> she is a sensible girl—which she is.[44]

Put to proper use, her skills guaranteed the social order desired
by males. An exception was the domain of male language, where the
telephone girl was an impediment to the male fraternity. If she were
unable immediately to discharge an impatient request, "the man who
is in a hurry swears softly to himself, forgetting that he is near the
transmitter." Such transgressions resulted often enough in fines, or, if
the culprit persisted, the withdrawal of the instrument by the phone
company.[45] Male expectations of both linguistic freedom and efficiency yielded to the delicate sensibilities of women, whose technical
clumsiness was the physical equivalent of moral unworldliness.

The telephone girl was generally not so fragile, and more often
depicted as a woman of ambiguous social status. Though frequently
in need of protection from predatory males, she was also bound to be
at their mercy by the service nature of her work. On the other hand,
she was independently employed, saucy in her pursuit of the slightly
racy recreations of the young and unobligated, and possessor of a free-
floating social identity that was particularly suspicious in women. In
short, she was in need of control. Her voice, symbol of both her work
and her gender, was the handiest extension of her for that purpose. "A
gentleman of fine ear, who uses the telephone frequently, suggests to
us that it would be a good thing to give the exchange operators a few
lessons in elocution, so that they might reply to calls with less nasality,
shrillness and snappiness of utterance," cautioned *Electrical World* in
1885, doubting that the class of women employed could speak correctly, or up to the standards of middle-class subscribers.[46] Such lessons might have the additionally desirable moral effect of enticing vulnerable operators from that "special detestation . . . the attractive skating
rink."

If working women managed not to transfer inappropriate oral models
to electrical communication or to make ignorant or careless mistakes

as telegraph and telephone operators, their decision to enter the world of electrical technology was sure to disappoint them in some other way—unless they were rescued in time to return to their appropriate role outside it. In the early nineties, a platonic friendship between a telegraph operator stationed at Banning, California, and another at the small desert outpost of Yuma, Arizona, blossomed into a romance when the Yuma operator fell ill and the Banning operator arrived by train to nurse him back to health with traditional female skills. "I, like a fool, had always taken it for granted that she was a man," the male half of the drama and the voice of the story explained. Marriage followed, and the Yuma operator's comment: "The Southern Pacific has lost an operator, but I calculate that I am ahead on the deal."[47] Loss of love was an occupational hazard for less fortunate women. An English version of the French play *La Demoiselle du Téléphone* turned on the fantasy of "a telephone girl in the execution of her duties overhearing her lover making an appointment with a music hall 'artiste.' "[48]

The drama of women's place on the stage of men's technology was constructed and reconstructed as consistently in electrical journals as elsewhere in society. Much of the romantic poetry featured as light filler in electrical journals metaphorically identified women with technological objects, both of them properly under male control. Graceful tributes flattered women to assert male dominance, in marked contrast to cruder displays of verbal or physical force that kept in line other underclasses, less likely to cohabit with men and requiring a different strategy of control. Called upon at a Minneapolis meeting of the National Telephone Association to acknowledge the ladies escorted by the male membership, W. H. Eustis, a prominent Minnesota lawyer, telegraphic entrepreneur, politician, and philanthropist, lavishly praised "woman the perfect telephone, the gift of gods to man." Both woman and the telephone were "inventions" second only to man himself. Sent down to please man, both woman and the telephone were mistaken for toys and turned out to be necessities. Just as a man filed a caveat and then a patent on his invention, "So when a man becomes interested in one of the fairest of American belles he becomes 'engaged' or 'files his caveat,' and 'serves notice' on all the rest of the fellows to 'hands off.' By and by the priest gives him his 'patent' and then he thinks he is all right for life."[49]

Endless stories of women's unpreparedness and incapacity in a world of technical expertise time and again demonstrated the reassuring conclusion that women would always depend on male prowess to conquer the world for them, however irritating their ignorance as the price of male mastery. The achievement by women of technological power, however modest, was shown repeatedly to have gone astray. Electrical journals depicted a stable sexual social structure in an otherwise uncertain, competitive world in which expert men might expect to bear the more difficult burden, but also the greater privilege of power, for a long time to come.

Endless variations on women's capacity to disorder a mode of communication thought to be ordered by an ineluctable natural law that males observed and enforced did have complementary comic relief in stories about nonexpert males who were befuddled by electric communication. Unlike expert men, they had no special information to communicate by telephone. Unlike women, they had no reserves of small talk on which to draw. *Tit-Bits* printed a story in 1897 about two male friends who found the telephone puzzlingly superfluous:

"Halloa Fletch! Do you hear me?"

"Yes."

"This is Sid. Thought I'd call you up."

"Glad to hear from you, Sid. How are you?"

"First-rate. How's things?"

"Calooshus. What's new?"

"Oh, nothing especially. Hadn't anything to do, you know, and thought I'd call you up."

(Pause.)

"Yes." (Another pause.) "Everything going on about as usual in the old town?"

"Yes, about as usual." (Pause.) "Awfully warm up here to-day. What kind of weather are you having?"

"Fine. Splendid weather."

(Pause.)

"Get the letter I wrote to you the other day?"

"Why, yes. Don't you remember I answered it?"

"So you did. I forgot." (Pause.) "Do you have any trouble hearing me?"

"Not a bit. Can you hear what I say?"

"Oh, yes." (Pause.)

"Well, how are you getting along?"

"First-rate. Anything—er—new going on?"

"No. Things are about as usual. It's—h'm—beastly warm here. Weather's fine where you are, is it?"

"Splendid."

(Pause.)

"Well, I must be going now. Awfully glad to have had a chance
to talk to you, old fellow."
"Glad you called me up."
"Good-bye!"
"Good-bye!!"[50]

Electrical Deception and Coercion

A proud and public component of professional identity was the integrity of the electrician who served no master but truth. Earnest stories
of exceptional personal and professional honesty abounded in electrical
journals. This important theme was rarely challenged, for intentional
deception by professionals charged with responsibility for complex
technical systems could imperil both human safety and public trust in
the expert knowledge on which that safety rested. The belief that orderly nature would exact swift and unerring retribution from any electrician who ignorantly misjudged or arrogantly misrepresented his expertise was thought to guarantee professional probity. Electricians
disciplined by science, it was claimed, could not be misled by personal
or political motives. On the contrary, the lofty standards of their
profession endowed them with general moral authority in human affairs. In 1898 E. G. Prout expounded on this theme to the newest
graduates of Stevens Institute of Technology as they prepared to tackle
the world's tasks:

> For some generations . . . natural depravity has been left to ministers,
> lawyers, editors, teachers, the mothers of families, to anyone, in fact,
> but the engineer; and this is where society makes a mistake. The best
> corrector of human depravity is the engineer. . . . Nature, calm and
> unrelenting, always stands looking at him. No other man in the world
> has such stern and unceasing discipline, and so it comes about that no
> other man is so safe a moral guide as the engineer, with his passion for
> truth and his faculty of thinking straight.[51]

Though experts appealed to the purity of professional integrity to
justify their claim to public trust, they did not feel bound to exercise
that integrity in their relations with stigmatized groups. Nor were they
concerned about the contradiction this posed to their claims of scru-
pulous professional honesty. Unselfconsciously reported instances of
deception and intimidation were treated as humorous and even praise-
worthy when practiced by experts on outsiders, but were outrageous
and intolerable impertinences when exercised in the opposite direction.

The more alien a particular technologically unempowered group seemed to electrical experts, the more blatantly coercion and deception could be exercised over it with the tools of electrical knowledge. As race, class, and station converged between experts and the technologically nonconversant in stories of their encounters, coercion and deception were less and less prominently featured. The general tenor of these stories nevertheless reflects the strain of class relations between electricians and their less technologically sophisticated social peers, subordinates, and superiors. In formula stories, marginal and despised groups were the focus of humiliating tricks by more powerful experts. The *Electrical Review* reported that Harry B. Cox of Cincinnati and his brother, an Episcopal clergyman "having charge of the city missions," had made "amusing" tests with an electric speaking trumpet devised by Harry Cox:

> They experimented upon an old darky, and completely frustrated the old fellow, who was walking up the road. Using the bell end of the horn, they began talking to the colored man as he walked along.
> The peculiarity of sound transmitted by the trumpet is that, to the person hearing it, it appears to come from some one near him, and not from a distance. The old darky hearing the voice was at first annoyed, then puzzled, and finally so badly frightened that he started up the road on the dead run, no doubt attributing his adventure to some supernatural agency.[52]

Thomas Edison was said to have startled a guest in his home, presumably a social peer, with a phonographic clock that announced the time to the unsuspecting visitor at 11:00 P.M., and the next hour called out, "The hour of midnight has arrived! Prepare to die."[53] A story in *Electrical World* touched on the tension between the worlds of practical and high culture. A French tenor visiting a telephone exhibition with an "electrical friend" was persuaded that he had been telephonically connected to his boss, the director of the Opera. Posing as the director, from the next cubicle the friend insulted the tenor's voice and threatened to halve his salary. The terrified tenor rushed out to confront the director in person, and counteroffered a one-third cut before the joke was revealed.[54]

In 1897 the *American Electrician* published a series of stories about electrically improvised pranks. "It requires practical experience in such matters . . . to appreciate them fully," one of the authors wrote, transforming the hostility these pranks expressed toward their irritating but powerless victims into a sophisticated mark of membership in the electrical fraternity. A stray dog was cured of his habit of stealing lunches by a charged wire baited with a juicy piece of meat.[55] At an electric plant, engineers wired knobs on doors and cupboards so that a full turn would give a shock to street urchins prowling "where they had no business whatever." Another station was troubled by youngsters who passed the time looking in the windows at "inconsistent and unreasonable hours," and who stood with their fingers hooked to the netting that prevented their entry, but not their persistent observation. Annoyed station operators electrified the netting with a charge strong enough to keep the youngsters from yanking their hands loose until the current was turned off. A portion of this story was devoted to a detached discussion of the risk of electrocution from this trick improperly done.[56]

The barbarity of these schemes made an impression on a few readers. A correspondent from Attleboro, Massachusetts, preferred electrical sport "with men who are my equals, and hav[e] more of a chance to get even than innocent animals and children." *His* target was dozing night watchmen. He devised an electrical apparatus to catapult a pail full of bolts, tin cans, and stones from the rafters near an unsuspecting guard at precisely the moment a strategically located alarm clock went off.[57]

Not all pranks were performed on lower-status victims, but those inflicted on victims of higher status were often more considerate, with less physical pain and fewer unpleasant surprises.[58] Generally, the humbler the station of a prankster relative to his victim, the less outrageous the deception permitted him. The nasty prank attributed to Edison, a figure of exceptional status, stood in sharp contrast to milder deceptions practiced by telegraph operators on a fickle public whose good will they required. Poking fun at those with no access to the special code of which they were masters was a way operators cemented the bonds of comradeship and laughed at the public without attracting its ire, unlike the hapless victims of electricians' pranks, whose resentment had little effect. In 1888, a scene from a play called *Across the Continent* offered an opportunity of this kind to the local telegraph fraternity. The leading actor played a telegraph operator at a railroad depot besieged by Indians. The dots and dashes he simulated were meaningless, but the audience accepted them dramatically as an electrical summons for assistance, answered offstage by another series of simulated Morse clicks. On one occasion, an authentic telegrapher manned the offstage sounder, and ticked off: " 'Say, Oliver, let's take a drink.' Which was received by 'Oliver' with: 'Thank God! We are saved!' "[59]

A proud and public component of professional identity was the integrity and honesty of the electrician, who served no master but truth. Intentional deception by professionals charged with significant responsibility for the social infrastructure could imperil critical human safety, not to mention public trust in the expert knowledge on which that safety rested, and which provided its symbolic authority. The belief that orderly nature would swiftly humiliate any electrician arrogant enough to misrepresent it was thought to be a reliable guarantee of professional probity. Electricians disciplined by science could never be misled by personal or political motives. Though experts often appealed to the purity of their professional honesty to justify their claim to public recognition, they did not feel bound to practice the same honesty in their relations with stigmatized groups. Instances of deception and intimidation that were reported unselfconsciously and described as agreeably humorous, even morally essential when practiced by experts on these groups, were considered an outrageous and intolerable impertinence of the underclass when practiced in the other direction. When some "smarty" telephoned the *Detroit Journal* with a hoax scoop of a fight aboard a vessel at Amherstburg, the *Electrical Review* printed an indignant account of the fraud written by the *Journal*'s editors: "As a hoax . . . it was neither ingenious nor clever. That is to say, it was a brutal lie." Urging the telephone company to take strong steps against future abuse, the editors lamented that so many innocents should be "at the mercy of any trickster or scoundrel that may place himself at the other end of the wire."[60]

A markedly double standard of professional honesty applied in the use of electricity to communicate with "savage" cultures. In the late nineteenth century of British imperial diplomacy, the expert electrician was a hero whose art subdued colonial troublemakers geographic and social worlds away. The famous electrical scientist and entrepreneur Werner von Siemens "found it necessary to intimidate the natives" while building the Djulfa-Tabriz portion of the Indo-European telegraph from London to Calcutta during the 1860s. Taking advantage of rainy-season conditions,

> he brought about a gathering of the natives and persuaded one of their notables to ascend a ladder and touch the wire, saying the wire would defend itself. On doing so, the man received such a shock that he fell down the ladder, and the wire was considered after that by the natives as being bewitched.[61]

The explorer Henry M. Stanley was accused of duplicity by "petty detractors" when it was learned that he habitually wore a concealed battery to deliver a mild electrical shock to African chieftains with whom he might shake hands. The purpose of this trick was to awe the native rulers with his "supernatural potency." This was not deception, insisted C. J. H. Woodbury, a prominent industrial engineer and spokesman who rose to Stanley's defense.

> It is beyond understanding why fault should be found with this harmless and efficient method of teaching a truth. The explorer received the identical physical effects which were imparted to the savage, the only difference being the mentality of the two races, the fact being that one was enabled by generations of civilization to ascribe the effects to material causes, while the other in his ignorance had nothing in his mental scope to apply a train of reasoning and, therefore, used the resource which ignorance always applies to the unknown, and supernaturalism was called in to fill the logical vacuum in the savage mind.[62]

Woodbury related another "lesson" from an incident in which a Plains Indian was electrocuted as he shimmied up a telegraph pole and chopped the wire with his tomahawk at precisely the instant it was struck by lightning several miles away. The point was not lost on the Indians assembled below, though the moral Woodbury drew for his expert audience had a different emphasis: "It is much better to use a savage to complete a circuit, than to make him serve as a target for projectiles, and the objection to this application of science for conquest is certainly more nice than wise."

Electrical World picked up a report from *La Lumière Électrique* of a "triumph of science over superstition" during British attempts to suppress the rebel leader Mahdi and his "wild followers" in the Sudan in 1884. A surprise attack at night was expected from the rebels and their Arab allies surrounding the town of Suakin, which the British held with the indifferent assistance of the "timid Egyptian soldiery, over whom the English officers had little control." The British prepared to defend their position by mounting great electric lights on towers overlooking the plain across which the attack was anticipated.

> Darkness fell, and the Arabs came on in hordes, shouting when they arrived within a few hundred yards of the walls, firing their guns at random, and waving their spears defiantly. At the right moment, the electric light plant was put in motion, the long beams of dazzling white light shot out suddenly upon the howling, rushing mass of Arabs, and in a few seconds the attack had by this means been turned into one of the strangest routs imaginable.[63]

A significant story element was the Arab failure to offer effective resistance. The British saw only random disarray as men fired guns in no apparent order and gestured wildly with crude weapons, and the futility of resistance to a superior technological, therefore civilized, order. A more subtle theme was the superior morality attributed to civilized communication over savage violence, as though the message from the British were not terrifyingly one-way, and backed up by force.

The efficacy of British terror in the Sudan was presented as a product of both scale and rationality. British lights had found their mark at a precisely synchronized technological rather than mythologically propitious moment. In a similar way, an account by Thomas Stevens recalled the "marvellous symmetry" of rows and rows of telegraph poles placed across the Persian plains "as evenly and perpendicularly as they might have been in Hyde Park." The English, he explained, "always take particular pains to have everything of this kind very superior in the East; it is a perpetual source of wonder and admiration to the natives, a standing advertisement of England's wealth, power and ability to the multitude who have no other way of learning."[64]

Criminals were another disenfranchised group over whom electrical intimidation could be exercised without qualm. Coercion in defense of the social order was the theme of a French report that electric lights had been placed in the Paris Morgue in 1888 "with the idea of increasing the effect produced upon murderers upon being confronted with their victims. Under the effect of the lights the 'confrontations' are expected to be much more effective."[65] Lights were also a deterrent to criminal activity, a substitute for more explicit control by authorities in Jacksonville, Florida, in 1895:

> If there ever was a time the city needed fewer policemen than have been necessary in the past, that time is now, since the brilliant illumination of the city by electric lights. Thieves hate light, and thugs despise it, and as a result (which the police annals will prove) there has been less thievery, less burglary and less thuggery in the city of Jacksonville since the city was lit by electricity than there had been in almost any corresponding period of the city's history.[66]

In the United States, professional electrical literature cast American Indians and Negroes as virtual members of a criminal class. The perplexity of these groups in the face of the white man's electrical machines made them easily manipulable, as when several horses were stolen in Julian, California, and suspicion fell on a local Indian.

Some one having introduced a telephone up there, the same was being exhibited, when it occurred to the owner of the stolen horses to get the Indian to come in and hear the "Great Spirit" talk. The Indian took one of the cups and was thrilled with astonishment at being apparently so near the Great Keeper of the happy hunting grounds. After some little time spent in wonderment, the Indian was solemnly commanded by the Great Spirit to "give up those stolen horses!" Dropping the cup as if he had been shot, the Indian immediately confessed to having stolen the horses, and tremblingly promised if his life was spared he would restore the "caballos" at once, and he did so.[67]

A particularly effective technique for flaunting electrical technology was to create with it something of value to the technologically unsophisticated (the voice signifying the presence of the god, in this case) more impressively than the technologically unsophisticated could create that thing themselves. The technological seducer's next move was to unveil the iron fist in the velvet glove. We will see this tactic used again in campaigns by scientists and engineers to discredit practitioners of magic and the occult, their nearest competitors for public interest and loyalty.

Four years after the Julian incident, a group of Apaches gathered before a telephone in St. Louis to hear another mysterious voice speak. According to the white narrator of the story, the claims these fierce warriors made to honor among their own people were based on brutal and daring exploits in battle. Primitive courage in savage exploits had not, it was noted with satisfaction, prepared them for the device white men regarded as evidence of the superiority of civilized science. The adversary culture was described:

> They were the leaders of the most implacable of savage tribes. Their hands had often been wet with the blood of murdered men and women; war-whoops of their tribe were as familiar to their ears as the cry of the wild wolf; but that intangible small voice which came to their ears from the infinite, that was a new experience to them.

The adversary culture was observed:

> One by one they listened to it; then, in silence, wrapping their blankets around them, they sat down to think. After a while their tongues were loosened, and each gave his idea of what the voice in the telephone was. The final conclusion was that it was the white man's Great Spirit, as he talked in English, and their anxiety was to find the instrument through which the Indian's Great Spirit spoke to his children.[68]

What interested the Indians was less the esoteric apparatus that fascinated their textually literate white hosts than the character of the voice they heard. In spite of the white man's contempt for the customs of traditional oral societies in the exchange of social knowledge, the conclusion reached by the Indians was astute. They had indeed heard the white man's Great Spirit. It was technology, the cause of the Apaches' anxious wish, doomed to disappointment, to find a voice of equal power of their own.

High Science and High Culture

The Perils of Popular Science

Priestly groups effect and maintain power by possessing significant cultural secrets. Training in the codes and rituals of these secrets is characteristically arduous, often lengthy, and reserved to elites. Along these lines, the restricted literacy of theoretical science and applied engineering was touted in the late nineteenth century as the exclusive property and singular responsibility of professional experts. Mastery of technical secrets was both an indicator of status and a path to it, conspicuously marked by ordeals of apprenticeship for select aspirants. Because of the esteem conferred on technological literacy by a society that reverenced it as a high secret, professionals were anxious to guard it from eager nonspecialists who might dilute it or, perhaps more alarming, possess it independently of the elites whose exclusive domain it was supposed to be.

Instant exceptions could be made for those who posed no threat to the status of electricians, especially if they lent their own prestige. John Jacob Astor's interest in developing an electric launch was praised for its promise of future projects funded by "a progressive gentleman of wealth." Astor was warmly described as "an enthusiastic devotee of the Goddess of Electricity for a number of years." It was not often, wrote a fawning electrical journal, "that a gentleman of Mr. Astor's position in the world devotes any time to electrical study," but his effort no doubt would be amply repaid in personal satisfaction.[69]

The enthusiasm of other nonspecialists was much in evidence, but not so gladly remarked. Public libraries and newspapers were the chief institutions of this enthusiasm. Electrical subjects were in high demand on the shelves of free libraries. At England's Brentford Public Library in 1892, Professor Silvanus P. Thompson's book, *Elementary Lessons in Electricity and Magnetism,* destined to go through forty editions in his lifetime, was said to be one of the 22 most popular books in a collection of 4,902 volumes.[70] The British journal *Lightning* noted that the librarian of the People's Palace in England mentioned electricity as "one of the pet subjects with its main readers," and furnished an "odd little list" of all the most popular authors on the subject, including Jenkins, Angell, Payser, Urbanitsky, Thompson, Munro, and Fergus. "The mixture reminds one rather of the man who said that his favourite authors were Anon and Shakespeare."[71]

In the columns of their journals, electricians deplored the belief that popularly available knowledge provided a point of entry into their field, something for aspirants to dabble in before setting up in business around some dubious invention. The triumph of journalism over proper professional training deceived both the public and aspiring electricians, with disastrous effects spelled out by the *Telegraphist* of London:

> Then there is the youth who has read as much as he could understand of an elementary textbook, and who has constructed a rude electrophorus out of a dripping-tin and a cake of resin, a galvanometer by coiling a few turns of wire round a compass box, and sundry other primitive machines. This youth is the genius of the family. He has a soul above quill-driving. He must be an electrical engineer. Taking advantage of sudden demand for men with electrical knowledge, he manages to get a situation, being ready with his set phrases, in which volts, ohms, and amperes are plentifully besprinkled. We next find him in charge of a dynamo, and shortly after read the account of his death caused by shunting some of the current into his own body. The inquest follows, and the father of the genius confesses that his son was very fond of reading electrical books, but that he had never had a proper training. To read about the potent force is harmless recreation, but to play with it sometimes means death.[72]

Pseudoelectricians thinking to bypass a certified technical degree were the continuing bane of the profession. "They are electrically inclined because they are good for nothing else," explained the *Electrical Review,* which also ridiculed their false textual training:

> They are the class from whom one, once in a while, gets a letter asking for the name of some book that "tells all about electricity, as I haven't time to read a library through." They become what after a time the telephone boys call P.L. electricians. These public library electricians are wise beyond their day and generation, but it is in the wrong direction.[73]

The P.L. electricians had got the point that professionalism was connected to texts, but they had not perceived the subtle and not so subtle ways in which proper electrical professionals and their texts were marked as an elite rather than popular order, in which those who earned the right to add "E.E." to their names, for "electrical engineer," would value the title that carried "so much dignity and authority . . . as they [did] their personal reputations."[74]

Newspapers and popular journals were as villainous as libraries. At a meeting of the Association for the Advancement of Science in New York in 1900, its president, Dr. Robert S. Woodward, lamented that "the elementary teaching and the popular exposition of science have fallen, unluckily, into the keeping largely of those who can not rise above the level of a purely literary level of phenomena."[75] Popular science education had failed because educational standards had been usurped by a popular press with the wrong textual skills and intellectual orientation for this function. As a result, "untrained minds fall an easy prey to the tricks of the magazine romancer or to the schemes of the perpetual motion promoter." Special ire was reserved for "schools of telegraphy," a catch term for all organizations that offered fraudulent electrical training. Regarding the crackpot theories of a "Kentucky colored preacher" who claimed that the earth was in danger of exploding from increased lightning production, and a "physician" in dread of "charging" the atmosphere with the steam and smoke of urban civilization, the *Electrical Review* editorialized, "these two scientists might be able, with the apparent knowledge they have of nature's forces, to open a school of telegraphy, somewhere."[76]

None of this was the path to true technological literacy. Experts believed that laymen were not only misled by popular scientific writing, but antitextual in general, or at least less rigorously textual than expert standards required. "In the wild dance that ushered in the electric light, the motor, and the box of electricity," a reader wrote to the *Electrical Review*, "sobriety of speech was at a discount and . . . every fellow who could invent and tell a good story of electrical feats that never occurred, pushed the student and thinker aside and took the front seat."[77] In expert culture, popular forms of knowledge such as telling tales—indeed, oral forms in general—were at war with the proper, or restricted, practices of scientific textuality. Popular standards of oral explanation could not convey the complexity or truth of science. Unfortunately, the man in the street read only occasionally, and then without diligence or discrimination. According to the *Electrician*:

It is sometimes said that a scientific theory is not thoroughly and clearly apprehended until it is capable of being explained to the man in the street. It is a hard saying, for there is no created thing more hopelessly out of harmony with the very atmosphere and foundation of science than that same street-walking gentleman. Perhaps he is not the same in all countries, but in our own country . . . no trace of the scientific spirit seems able to permeate the mind of the average man. His interest in science begins with an experience of its application to . . . pecuniary profit, and ends with the perusal of some very light popular illustrated text-book.[78]

A true and full appreciation of scientific knowledge was off limits to all except properly schooled experts who belonged to restricted textual communities. The lofty *Electrician,* for example, despised popular science, and its editors were quite unconvinced that ordinary men could profitably share in scientific knowledge. It divided the domain of scientific writing into "pure" science, which traced out truth, "applied" science, which appealed to the pocket, and "popular" science, which endeavored to render truth "in a pleasing form."[79] Popular science was a sport for immature and gullible intellects, a cheap imitation of the form of scientific textuality without its substance. Consider these observations from the *Illustrated London News* on the characteristically brief, unskeptical presentation of the "facts" of science in popular periodicals:

It is conveyed in brief but most attractive portions, and never hampered with details as to how the intelligence was procured, the instructions drawn, or the figures calculated. In the present issue of my favourite periodical, I read that "Icebergs last for two hundred years." One cannot help wondering how this information, doubtless gathered from trustworthy sources, has been obtained. Do you catch your iceberg young, and watch its growth, deputing the interesting task to your descendants, or do you select one from its companions on account of its vast proportions, and note from decade to decade its gradual diminution? "It takes a snail fourteen days, five hours, exactly, to travel a mile." What patience and assiduity it must have taken to record this fact with accuracy! How curious, too, to discover that all snails have the same rate of progress! . . . "Persian women have a horror of red hair." How few of us are acquainted with Persian women, or could have learnt this by other means! How enterprising must be the periodical which sends, perhaps, a special correspondent to ascertain such a circumstance. . . . No work of information has ever given me the pleasure I derive from these weekly additions to knowledge. Sometimes they surprise as well as delight me, for example: "Kissing originated in England." Heavens![80]

True technological literacy for electrical engineers could be learned only at proper professional schools, the kind Henry Floy wrote about in an *Electrical World* article published in 1894. Floy was a true man of science, a Cornell graduate in mechanical engineering whose professional reputation was based on his work on long-distance power transmission. Floy divided electrical engineering aspirants into "students" at classical or technical colleges and "artisans" working their way up from the practical end of the profession. He portrayed a field crowded with all the electricians society could use, unable to absorb many more in the near future. In this struggle the competitive edge belonged to restricted textuality, to students equipped with high technological literacy instead of to craft artisans, throwbacks to a preprofessional age. "The college curriculum cannot help but make [the students] exact thinkers in addition to furnishing them a supply of theoretical knowledge which the 'practical man' has failed to obtain." Whatever temporary advantage artisans might have over inexperienced graduates would vanish as soon as the latter acquired a modicum of practical experience. "It is but natural," wrote Floy, "that they should outstrip men who have not received the advantages of a college training, and in very many cases, not even a high school education."[81]

Who did not understand what it meant to be a professional, expertly conversant with esoteric technical literature? Who had no reliable knowledge of the need for electrical engineers, or the training they required? Who thought to bypass the stringent, textually oriented selection procedures built into professional training? Who else but the popular, that is, the nonprofessional press. The existence of a pool of aspirants without prospects was laid at the door of the "present depression . . . but more truly to the overpopularizing of this particular profession," which brought too many hopefuls flocking. Driving home the point that professional training was the prerogative of elite institutions, Floy offered catalogue descriptions of the elaborately technical electrical engineering course at Princeton, a "classical" school, and Cornell, a "high-grade technical school." He cited figures to prove that too many students were being trained even in these schools, warning that most would be forced into low-paid jobs beneath their expectations unless they were exceptionally talented or well connected.

In the sight of experts, the popular press erred equally in its worshipful admiration of all things electrical, and the wizards who brought them into being, and in its wild and immoderate extrapolations from the reasonable fears experts sought to instill in laymen to keep them physically and intellectually at arm's length from electricity. Though preferable to criticism, worship made experts nervous because it raised the specter of popular dissatisfaction with electrical utopias expected and not delivered. Too diffuse a fear of electricity, on the other hand, could damage expert interests. A characteristic mixture of concern and contempt for popular misgiving was expressed by C. C. Haskins, the president of the Wisconsin Telephone Company:

> That thousand-tongued old gossip, Madame Rumor, will never lose time nor opportunity to charge the "death-dealing" fluid, as she calls it, with anything in the way of a casualty which occurs within gunshot of a dynamo or as near as forty rods from an electric lamp or motor, provided the enterprising newspaper commissary can dish up distorted facts enough to give the gossips a good square meal. Now and again when I see exaggerated accounts of some trivial affair connected with electricity . . . I am reminded of the wholesale cackling in a barnyard, when some silly pullet has accidentally run upon a toad in the fence corner.[82]

That Rumor was female and her medium oral, and that she was in league with journalists, was no accident of characterization.

Perhaps more distressing to experts was that popular authors, readers, and lecture audiences were generally oblivious to criticism. Experts lamented that popular science writing did not take seriously its responsibility to enlighten its readership. Their concern was perhaps less for the truth than for the perception that popular writers did not sufficiently school laymen in obedient submission to expert authority. Experts who complained that popular lightheartedness toward electrical science was either ill-concealed arrogance or exaggerated expectation were confronted with a predicament, since it was essential for experts to court the public in order to achieve a status based in public perceptions of merit and to secure popular acquiescence to expert judgment in matters of electrical interest. Except for the purpose of flattery, however, anecdotes in the expert literature revealed scant desire on the part of electricians to share their accumulating power with the public on whose behalf their privileges were justified. The public was that group in whom enough interest had to be shown to ensure its support, but with whom caution must be used not to promise too much.

The tireless and demanding public was ever present. At the seventh annual convention of the National Telephone Exchange Association in 1885, Angus S. Hibbard, then superintendent of the Wisconsin Telephone Company, recalled the flowering of public expectation following the "first crude experiments" in telephony. "The usual newspaper phraseology" shouldered much of the blame:

> It was almost suggested that the life of the average American would be incomplete were he to omit from his daily routine the pleasure of telephoning to his friends in Japan.
>
> Our friends in New York, over their breakfast coffee, were to have from London oral cognizance of what was most thoroughly "English," on that same afternoon, echoed—at the turn of a switch—by the midnight songs of the seals at the Golden Gate. . . . The wide-spread publication of imaginative successes has, in deceiving the public, robbed the laborer in this field of any laurels he might hope to obtain, by making his accomplishments appear tame in view of the marvels advertised for his art.[83]

Expert appeals for popular support often implied that universal electrical prosperity was not far off, especially for groups that had not been visible beneficiaries of industrialization. Perhaps so as not to assume too willingly the obligations of that promise, many of these appeals linked electrical progress to the necessary contributions of all who possessed useful skills and ideas. In the technological society of the future, all talents would be impartially judged and rewarded. Literally construed, this appeal made expert exclusivity hypocritical, and fomented ideological instabilities of which the emerging expert elite was uneasily aware. The more glowingly the public consented to the efficacy of expertise and desired some demonstration of its largesse, the more firmly were the barriers raised against it.

Despite the best efforts of electrical professionals to keep the crowd from the door, occasionally one or another got in. But where the outsider had penetrated the circle and could not be expelled, vigilance against intrusion was matched by other strategies to assimilate him and make him an insider. Within the professional electrical community, Guglielmo Marconi's introduction of wireless telegraphy in England in 1896 provides a particularly clear example. Marconi was not a product of the British scientific establishment, or even a Britisher, but a youthful Italian, an amateur experimenter whose claims threatened to make publicly and professionally ridiculous several prominent British electricians and scientists. He was, in other words, a threatening outsider.

The expert press at first reacted suspiciously. Marconi was accused of having done nothing that British scientists had not done already. A British journal sniffed that at the Royal Institution, where he was presented by his mentor, William H. Preece, chief engineer of the British Post Office, he was addressed as "Signor Marconi," since he "has hardly been in this country long enough to entitle him to be referred to as 'Mr. Marconi,' though Mr. Preece does so."[84] Worse, Marconi had gained the lavish favor of the popular press. "It is a pity that Signor Marconi should choose to give his information first to the lay press," complained the expert press, "and thereby insure that his grain of wheat should be hidden in a bushel of chaff."[85]

When the importance of Marconi's achievement finally could not be denied, efforts were made to efface the damaging accusation of outsider, and almost overnight Marconi became an insider. In 1899, he was described as follows in the *Western Electrician:*

> Although Italian on his father's side and by birth, Marconi's mother was an Englishwoman, and in general appearance, complexion and manner the inventor is seemingly a young and clear-complexioned Englishman. Signor Marconi is a man of medium height and slender, with blue eyes and clean-shaven face, except for a slight mustache. He is of a retiring disposition. As might be expected, he speaks English like a native. The Herald described him most aptly: "Signor Marconi looks like the student all over and possesses the peculiar semi-abstracted air that characterizes men who devote their days to study and scientific experiment."[86]

Controlling Language

Conflict between exclusive certification and diffusive popularity turned on the more basic debate about whether the skills of electricians were really practical or theoretical, manual or textual, craft or scientific. The aura of elitism enjoyed by professionals, with its canon of exclusivity, underlay the argument that sophisticated textual and linguistic skills were the mark of the professional electrician. Not from imitating skilled electrical craftsmen, but only from the diligent study and informed discussion of technical books and papers and professional journals, could the secrets of electrical science be learned. This made correct technical language correctly used essential to the expert's claim to professional authority.

In 1898 the great Elihu Thomson wrote a letter to the *Electrical World* concerning an error in a technical procedure reported in its pages by the young Reginald Fessenden. "I have to thank Professor Thomson for correcting me in the matter referred to," Fessenden replied. He pointed out that the standard written authority on the subject had given the same account. "The fact that even such a standard work as 'Watt's Dictionary' though devoting a half-page, Vol. III, p. 888, to iron amalgams, and describing three ways of making them does not mention this method, is, I believe, a sufficient excuse for my mistake."[87] A recommendation by William Crookes had led him to his original method, he continued. The young scientist defended himself by showing that he had consulted the proper experts and that his error had not exceeded that of the most authoritative text. Finally, he submitted graciously to the authority of an expert more credentialed than himself.

While this example of science at work had little homiletic value for a popular audience, electrical journals delighted in anecdotes about how technical knowledge was misunderstood and misapplied by amateurs, charlatans, and even students. Such anecdotes were part of the ritual for excluding charlatans and the inadequately schooled, while reminding legitimate community members how difficult the subject of their study was, even among the hard-working technical literati. "The young electrician asks for information," explained the *American Electrician* in 1898, "and finds that his misconceptions are so ludicrous that he gets roundly laughed at. To avoid ridicule, he forbears to ask again, and unless he dissolves his difficulties by strength of intellect, they remain a continuous puzzle and protest against the accuracy of science."[88] Perhaps the classic story in this vein was about the student who, asked to define electricity, said he used to know but had forgotten. "How sad," replied his weary professor, "the only man who knew what electricity was has forgotten."[89]

Popular literature that treated electricity with cheerful irreverence never failed to arouse the suspicions of the sober scientific and professional press. The *Electrical Review,* for example, fulminated at doubtful electrical gimmicks of plot devices in light fiction. Most authors "make absolutely incorrect use of electrical terms, a few seem to grasp the meaning of the simpler elements of science, while others deal with glittering generalities to avoid positive errors."[90] To convey accurate technical information was rarely the purpose of electrical terminology in popular fiction, of course. It did greater service by imparting an air of up-to-date excitement. It fulfilled, if only for a moment, the fantasy that the much admired electrical magic could be possessed by the simple act of calling its name.

A children's story in *Harper's Round Table* illustrates the point. A piece of light fiction, it appropriated scientific content, or something like it, for entertaining ends. Its central characters were Jimmieboy, a kind of Everychild, and an imp whose home was the telephone instead of a magic lamp. In the course of their adventures, the imp showed Jimmieboy a pushbutton-controlled electric dictionary containing all knowledge. The dictionary defined *battery* in an esoteric technical fashion.

> "Understand that, Jimmieboy?" queried the Imp, with a smile, turning the Dictionary button off.
> "No, I don't," said Jimmieboy. "But I suppose it is all right."

The Imp lectured on batteries and electricity in terms practical and accessible:

> ". . . a battery is a thing that looks like a row of jars full of preserves, but isn't, and when properly cared for and not allowed to freeze up, it makes electricity, which is a sort of red-hot invisible fluid that pricks your hands when you touch it, and makes them feel as if they were asleep if you keep hold of it for any length of time, and which carries messages over wires, makes horse-cars go without horses, lights a room better than gas, and is so like lightning that no man who has tried both can tell the difference between them."[91]

In popular stories, effects of announcements were "electrical,"[92] and ideas came to characters "like an electric shock."[93] Puns were widely used, since they were effective even when audiences grasped only dimly the technical language involved. "George, dear, what kind of fruit is borne by an electric light plant?" asked the *Terre Haute Express.* "Electric currents, of course."[94] "There's music in the electrical air," the *Electrical Review* prefaced another pun that satisfied its primmer standards of veridicality: "When the intelligence of a disaster is flashed over the wires electrically, a newspaper very properly calls it 'shocking' news."[95] London ladies "are becoming interested in our science," wrote the *Electrical Review,* "and are fast abandoning the afternoon sewing bees and 'high tea,' and taking up 'electricitee' instead."[96]

It was commonly observed that electricity had introduced a remarkable number of words into the language. The *San Francisco Chronicle* compared the fecundity of electrical language to the inventiveness of terms and discourse about patent medicines, which popular language and culture revered with a similarly expansive and uncritical enthusiasm.[97] New words permitted those who were richer in awareness and imagination than in educational or technical resources to have their own special entry into electrical culture. The process of selection by which some words and ideas were popularly embraced and others were rejected certainly had little to do with expert judgment. New words were like new inventions, explained *Cosmopolitan*. "If they prove popular, they hold their own, like a derby hat, or ice-cream soda, or electric lights or telephones."[98] Although experts used humorous, playful language to deny status to outsiders, and appropriated what they regarded as comical aspects of the dialects or vocabularies of outside groups to disparage them, they fretted, without noticing the contradiction, about whether popular play with technical language was morally sound. Playful expressions of lay scientific interest were disturbing to editors and readers of journals like the *Electrician,* since any linguistic breach in the dike of high science had unpredictable potential for diminishing expert claims and opening up electrical culture to those who lacked credentialed membership.

Stratifications of rank required demarcation not only between members of textual communities and outsiders, but also within textual communities. Pure scientists wished to be distinguished from applied engineers, and engineers wished to convince scientists, one another, and the public of their indispensability in the management of crucial social problems. Facility in manipulating technical symbols was therefore a jealously wielded instrumentality. Cadres of operatives also had to be coaxed to accept the delicate bargain in which they identified with the technological enterprise that required their participation, but were clearly set apart from those upon whom greater decision-making authority was conferred by dint of their superior command, among other things, of powerful technical language. To be a professional, and particularly to be authentically discriminable from the large class of pretenders to professionalism, required symbolic distinctions that experts believed only they could appreciate and, more to the point, enforce. Control of technical language was a means for experts to establish themselves as arbiters of the domain of technological reality and, from that strength, to seize the larger domain of social reality. Every bit as crucial as correct professional education, therefore, was the orderly invention and development of a technical language in which experts modeled and polished the world that belonged to them.

Technical language was not easy to control. Efforts to standardize and rationalize it on the model of the lawful universe it ostensibly described were continually frustrated by its expanding variety and complexity. "There is scarcely a single term employed in electrical literature upon which different writers thoroughly agree," commented *Electrical World* in 1885.[99] *Western Electrician* observed that the second edition of a dictionary of electrical terms by professor Edwin J. Houston, a well-known electric lighting entrepreneur, had five thousand entries and as many cross references. Published in 1892, it was completely revised from the original text that had appeared three years earlier.[100] The progress of new words and phrases was tracked with interest and often with controversy. Erratic and erroneous usage occasioned expressions of distress and flurries of proposals for formal regulation. Professional bodies like the International Society of Electricians and the English Society of Telegraph Engineers and Electricians appointed committees to consider what could be done about the lack of uniformity in electrical terms and usage.[101]

In 1893 Professor E. Hospitalier, a Frenchman, chaired a Committee on Notation for the International Electrical Congress, which recommended wide-ranging standards for electrical language and systems of notation. Hospitalier deplored the divergence of views and usage among electrical authors, which stemmed in his view from "the absolute lack of method" in scientific and electrical notation, the principal purpose of which, he declared, was "to enable engineers to understand the formulas published in the works of different countries."[102] The editors of *Electrical World* welcomed his system of forming compound words and naming units as being "logical throughout and based on correct principles," as well as internationally uniform.[103]

Other electricians relished the anarchy of countless small debates, complaints, and case-by-case proposals for specific usage that filled the pages of professional journals. In response to an inquiry about the best word to express "execution by electricity," the *Electrical Review* reported a variety of suggestions, including *elektrophon, electricize, electrony, electrophony, thanelectrize, thanelectricize, thanelectrisis, electromort, electrotasy, fulmen, electricide, electropoenize, electrothenese, electrocution, electroed, electrostrike,* "and finally joltacuss or voltacuss."[104]

The *Electrical Engineer* took a lively interest in such matters and framed the problem in these terms:

> The charge is made against this country that slang is too freely adopted here and that new words are welcomed simply becase they are novel, and despite their hybridity and barbarism. In some respects the charge is true, but with a new society, living under new conditions and ac-

quiring every day new implements of industry, how is the check on base word coinage to be applied and who is to enforce it? If the stock of old true coinage is sufficient, who will be capable of proving it to be so? And if the old coinage has to be melted down in order that new true coins answering to new wants may be furnished, who will do the minting?[105]

An instructive debate concerned an attempt by George W. Mansfield of Greenville, New Jersey, to introduce the word *motoneer* to designate "the man running the electric motor." In a fraternal letter to the editor of *Electrical World* in 1884, he cited precedents for this use in "several articles" and an "interesting little work," presumably popular, called "Wonders and Curiosities of the Railway."[106] Though in appealing to textual precedents Mansfield followed a form cherished by professionals, these particular precedents were unpersuasive. Indignant opposition reared its head in the letters column. In the view of Thomas D. Lockwood, an electrician for the American Bell Telephone Company, coiners of new technical terms already had much to answer for. Lockwood scorned *cablegram,* a word whose existence implied a need to dignify in separate words the existence of different proportions of submarine to land-line cables in message transmission, and *electrolier,* a pretentious analogue to *chandelier. Motoneer* he regarded as "the worst specimen yet launched upon a long-suffering public."[107]

Despite his claim to have popular interests at heart, Lockwood's arguments appealed to elite expertise and to the preservation of a social order based on it. The word *motor,* he argued, was inaccurately used already to describe "the engine by which the electric energy is utilized, and should only be applied to the electric energy so employed." If the attendants of electric engines could be motoneers, "the attendants of gas engines, steam engines, air engines, etc., are entitled to the same distinction." The articulated premise of his argument was that unnecessary words should be avoided; its tone took offense at the use of professional terms to describe the laboring classes.

Also put off was "Ampere," an anonymous correspondent who scorned *motoneer* as a dubious analogue to *engineer,* even taking into account the corruption of the latter term in everyday use. Originally, according to "Ampere," *engineer* had meant exclusively

"a person skilled in the principles and practice of engineering, either civil, military, mechanical, marine," etc. According to this definition the man who designs or builds an engine is an engineer. . . . In America, however (and I believe nowhere else), the meaning of the term has been extended so as to make it apply as well to the person who drives or manages an engine. Indeed to many persons in this country the word engineer possesses no higher meaning than the latter.[108]

In England a more orderly and less popular usage was current, he claimed. There, *engineer* retained its elite meaning and the terms *engine-man* and *engine-driver* covered the case for which *motoneer* was proposed. In a tactical appeal to high culture, "Ampere" quoted Macbeth ("Take any other form but that, and my firm nerves shall never tremble") and invoked class solidarity:

> The man who manages the switches seems perfectly satisfied to be called a "switchman;" the man who manages the brakes is content to be called a "brakeman;" the man who manages the boilers does not object to being called a "fireman," and the correct term for one who manages an engine is "engineman." Pray, then, why not stop inventing new words and adopt "motorman." It is short, and to the point. It speaks for itself. . . . It is, indeed, a perfectly legitimate word, with a long line of pure Saxon ancestry—tradesman, workman, footman, etc.—and a full line of relations, sisters, cousins and aunts, like cabman, hackman, boatman, etc.

While both protagonists claimed that only the loftiest considerations governed their objections to *motoneer,* their arguments made sense only within a framework of elite distinctions. The dike could be broken at no point. Engine operators must be verbally located in the sturdy "Saxon" class of tradesmen and laborers; elite terminology must be derived from Latin or French, and not the slightest linguistic hint should elevate operatives to the status of engineering professionals.

Builders of a future inherit a past that they reshape. To justify their claim on the future, electrical experts never tired of comparing themselves with cultural ancestors whose achievements they regarded as exactly parallel to their own. Experts, too, had created a priestly language and an authoritative textual tradition. They styled themselves as the continuing link in a cultural tradition charged with preserving Western civilization for future generations. Other elites had guarded and extended the textual canon of Western culture in specialist languages that made civilized life possible and invested it with its distinctive character in different eras. Talented electricians were likewise stewards of a trust to develop electrical science in specialist languages and texts best suited to that task. In expert eyes, the important difference between the great tradition and electrical science was historical need and occasion. "Those who spend their lives among the dreams of the ancients, knowing nothing of the powers and achievements of modern man, may be pardoned for proclaiming their own inferiority," wrote *Scientific American* in 1876, "but they have no call to speak for the real men of the real world about them, the men who are doing the world's work, at the same time steadily lifting humanity to higher and yet higher planes of capacity and power."[109] Many hoped that the humanist languages of the past, testament to grand but fading glories of previous human achievement, would yield to the more serviceable and progressive language of high science. "The fables and fairy tales of old pale before the facts of the present day," proclaimed President Wilmerding at the opening of the Electric Light Association convention in New York in 1896.[110] Students of the practical application of electricity, observed *The Telegraphist,* "find in it more real enjoyment than any puzzling over the vagaries of the modern poet or poring over the meaning of ancient cynics can afford."[111]

An incident on the occasion of the first annual meeting of the New York Electric Club in 1887 encapsulates these tensions between scientists and humanists, and the logic by which electrical experts aspired to equal status with culture heroes of the past and made their own bid for historical superiority, since the task of salvation was now theirs. The meeting's program featured a performance by club poet William "S(hakespeare)" Hine of his original and flowery work, "Electrical Nomenclature."[112] Borrowing themes from the Bible and classical mythology, Hine invented amusing parallels between electrical terminology and the elegant language of classical discourse.

With the self-deprecating modesty of one secure in his social station, but also in terms intended to suggest effortless familiarity with the canon of high culture, the poet described himself as being as

> well-trained as most
> (And I do not wish to boast)
> As a second Ananias in plain prose

and offered his thesis that

> Searching mythologic tome
> We find the cause from which our troubles flow.

The trouble in question was the proliferation of awkward, cacophonous, and obscure electrical terminology, language many found unworthy to succeed the dignified and graceful language of the great tradition. In explanation, the poet recalled the Promethean theft of fire that first brought punishment from the gods upon mankind. He reported that Jove the Thunderer had watched in alarm as mortals examined pieces of amber from some ancient Vesuvian eruption and discovered the properties of "Electron." Assembling the gods, Jove warned that mortal

> fame would wrest from us our mighty powers,
> And make the elements their servants like to ours,
> And seek to give the widest of publicity
> To all the hidden powers of electricity.

Reminding them, in a reckless mixture of traditions, that the Tower of Babel had once before thwarted mortal emulation of divine power, he urged the gods to

> try it on again; it worked well then,
> and kept us still above the herds of men.

And this was why, the poet claimed, the gods created the vocabulary of electricity expressly to confuse men:

> . . . the words came crowding thick and fast, ·
> And each more harsh and puzzling than the last.
> Watt and erg and dyne and ohm,
> Gauss and ampere and coulomb,
> Armature and commutator,
> Secondary generator,
> Inductorium, insulation,
> Permanent magnetization,
> Pole and arc and dynamometer,
> Field of force and electrometer,
> Anions, ions, cell and cation,
> Radiophony, amalgamation,
> Agonic, isocinic lines,
> Coils and shunts, circuits and sines,
> Rheotrope and galvanometer,
> Polarization and rheometer,
> Molecule and atomicity,
> Currents and cosmical electricity,
> Endosmose, osmose, diaphragm,
> Electrolysis, A. J. Dam,
> Amplitude and declination, ·
> Micro-tassimeter, retardation,
> Resinous, vitreous, chlorous, switch,
> Phlogiston, solenoids "and sich,"

Zincode, platinode and relay,
Calorimeter, inertia,
Diamagnetism, fusion,
Resistance, equivolt, occlusion,
Dielectrics, actinism,
Foucault, helix, crith and prism,
Nascent, tension, rheostat,
Equivalent, Leyden jar, Joulad,
Equipotential, consequent pole,
Just here they stopped—God bless my soul,
The gods themselves had e'en gone mad,
The scheme that they for others had
Contrived had done its fatal work,
Try as they might, they could not shirk
The fatal consequence; that's why
They quit the realms of earth and sky,
And sought out home in other spheres. . . .

Claiming that electrical inventors had taken upon themselves a task that had defeated even divine intelligence, he concluded:

Each and every inventor, both early and late,
Seems doomed, by an incomprehensible fate,
To follow serenely the deities' plan,
Of clothing each new scientific idea
In garments ungainly and ancient and queer,
Until what knocked the gods out proves fatal to man.

This fable of the origins of technical language made a clear point. Electricians believed the time was long past for a shift of cultural reference from the authority of the humanities to the authority of science. Despite their professed regrets about the unwieldiness of the textual torrent loosed on the world by science, they were as proud of it as if it were divine revelation, which in some sense they thought it was. Notwithstanding ritual genuflection to humanism on official occasions, they felt their own endeavors were far more likely than elaborate systems of false myth to unlock the secrets of the cosmos.

To demonstrate to laymen the superiority of scientific discourse even at play, electricians were forever recasting mythical language, when it intruded, in scientific terms. "Too many clever electricians," William Crookes once remarked in a discussion of the danger of alternating current, "have shared the fate of Tullus Hostilius, who, according to the Roman myth, incurred the wrath of Jove for practising magical arts, and was struck dead with a thunderbolt. In modern language, he was simply working with a high tension current, and inadvertently touching a live wire, got a fatal shock."[113]

Proofs of Priesthood: Electrical Magic

Although an express mission of science was to kill magic and myth, electrical experts were deeply implicated in the production of both. "One miracle has followed another until we can but wonder what apparent impossibility will be accomplished next," intoned the National Electric Light Association's President Wilmerding.[114] Scientific progress was produced by manipulating mysteries known to an informed priesthood and accepted on faith by an audience to whom it was demonstrated at regular intervals. This was in spite of a powerful self-justifying belief by electrical experts versed in restricted scientific textuality that they labored exclusively for an audience of peers, and ventured into the larger world only when they could do so without compromising scientific principles in the slightest. Claiming that their work benefited the world in ways the world could neither imagine nor truly understand, they resisted every outside suggestion that they explain themselves to the masses.

As a practical matter, this insistence on a zone of elite exclusivity was only relative. Pursuit of large-scale magic, even by influential and informed minorities, always requires a degree of support from a larger, less informed community able to seriously restrict the scope of the more elite enterprise if it becomes sufficiently restive. The need of scientists and engineers for public sympathy created, first, a popular spectator audience for whom science was marvelous entertainment on instrumental, intellectual, and aesthetic grounds; second, a scientific priesthood attentive to the value this audience placed on magic, though it regarded that fascination with ambivalence; and, third, a class of performers, also sensitive to the public's affection for magic, that stood ready to challenge this priesthood for the hearts of its following. The priesthood considered this last group to be charlatans, badly informed and entirely insincere, exploiting cheap dramatic effects and debasing knowledge for its own profit.

The *Electrical Review* sent its reporter to interview Professor Hermann, nightly performing "most wonderful" tricks "by the aid of the fluid" in the Temple of the Sphinx on Broadway in New York in 1892. Informed that Professor Hermann had constructed a watch that could generate one horsepower of electricity, the reporter was astonished to discover "that there was such a machine in existence and that the *Elec-*

trical Review had heard nothing about it."[115] To see the machine was forbidden, for it could not be replaced if lost or damaged, the professor's business manager explained. It was described for the reporter's benefit, reproducing the hollow form of a cardinal scientific rule of verifiability, if not, to the reporter's glee, its substance. "Inside of the case is an armature, some coils of wire, a few magnets, a lode stone—I suppose that you know what a lode stone is? . . . Then there are some push buttons, a dial with some numbers, some switches, and a few other things, and that's all I can remember."

It was said the professor also had invented a telephone that transmitted images. "For instance, in one of Hermann's séances, he holds a photograph at one end of the telephone while he stands in the orchestra, and a reproduction of the photograph is seen on the stage in midair." To the question of why the professor had neglected to market such a remarkable device, the business manager explained that this would rob the professor of the audience that flocked to his performances. This answer did not satisfy the reporter, who believed that the professor's motivation was economic rather than scientific. But it mimicked a scientific distrust of merely practical applications of the higher magic to which scientists were privy, as well as an appreciation of mysteries and miracles that exceeded the understanding of a less textually adept audience.

That scientists and electricians who were not tagged as charlatans also bent electrical knowledge to dramatic ends was deemed legitimate in the pursuit of truth. Instances included lectures with striking and wonderful "effects," and accounts of scientific work suspensefully constructed in written form for popular consumption. The "finale" of an electrical lecture by Edison Company representatives in Boston in 1887 was a spiritualistic séance. "Bells rung, drums beat, noises natural and unnatural were heard, a cabinet revolved and flashed fire, and a row of departed skulls came into view, and varied colored lights flashed from their eyes."[116] So impressive were these effects that a second lecture had to be repeated, for a standing-room-only audience. In 1894 the Sunday *World* attached this caption to a picture accompanying an article by Arthur Brisbane about a lecture by Nikola Tesla: "Showing the Inventor in the Effulgent Glory of Myriad Tongues of Electric Flame After He Has Saturated Himself with Electricity."[117] Magic and poetry cloaked in science satisfied the aesthetic impulse of scientists within the rules of a textual community that frowned on their more familiar popular expressions. To the scientist and engineer, magic meant not what *they* did, but nonscientific modes of discovering and transforming reality, and especially the lack of an intellectual discipline associated with that science and mediated through literate practices.

When audiences got the metaphorical point that experts were at some pains explicitly to deny, namely, that science was a magical enterprise superior to lesser forms of magic, they were inclined to push that point to its logical conclusion. This literalness embarrassed scientists and sometimes forced them to disavow their own propaganda. The line between debased and scientific "magic" was sufficiently ambiguous to be a constant concern. Comparing Tesla's theatrical appearance at the Royal Institution in London in 1894 with a more straightforward, less enthusiastically received demonstration of Hertzian waves by Oliver Lodge, the *Electrician* weighed the need to present scientific information dramatically enough to capture public interest against the competing desire to convey a dignified appearance that would guarantee the respect of the scientific community. The difficulty in this tug of war was keeping the credibility of the priesthood intact. "It is not, of course, desirable," the *Electrician* editorialized,

> that these Friday evening discourses should, as a rule be tinged with quite so much of the dark *séance* style, but . . . had it been impressively and reiteratedly brought home to the spectators last Friday evening that what they were witnessing in the experiment necessarily involving so much shouting but resulting in a minimum of sparking, was the transmission of electromagnetic waves right through three brick walls and across the staircase and the room adjoining the theatre, it is possible that even the most uninitiated would have left the Royal Institution that night with a tolerably concrete idea of the importance of the experiment.[118]

Even less clear was the status of electrical tricks practiced by magicians for aims approved by professional electricians on those occasions when both groups were allied with political authority against subject peoples or other underclasses. Following the conquest of Algeria, the French government commissioned Robert Houdin, a celebrated conjurer, to challenge the popular authority of the native Marabout priests, whose following the French coveted in order to execute their colonial program. Houdin's object was to "convince the Marabouts that a European could not only surpass their miraculous exhibitions, but expose their fraudulent occult demonstrations, with which they continually awed the more ignorant Arabs, and excited others to discontent and rebellion." Houdin devised "The Light and Heavy Chest," his greatest trick, to astonish and humiliate the native necromancers. He exhibited an apparatus that he claimed would give them pain or make them weak at will. Accounts of this contrivance differed, although the incident was a staple of discussions about electricity and the control of colonial people. The device seems to have been fitted with secret pulleys and electromagnets that thwarted all attempts to lift a special iron chest displayed with it. To finish the performance, Houdin administered electric shocks through this chest "to strike terror into the soul of the native."[119]

Such efforts to subdue exotic cultures drew no criticism from electrical professionals, although not all electrical strategies were as aggressive. A German traveler, Dr. Junker, visited the area around Khartoum between 1879 and 1883 without incident by carrying an assortment of picture books, masks, musical instruments, puppets, electric toys, and other things "of which the natives knew nothing," with which he entertained them in order to win their friendship.[120]

The occasional appearance of sensational stories about the occult at the fringes of scientific and professional literature was usually legitimized by the suggestion that science was absorbing and taking over disreputable magical modes and replacing them with benign scientific ones. Stories of this kind dramatized the heroic encounter with the unknown, and the contest of power against power. As in any thrilling story, the plot of scientific proof and verification revealed what was authentic and legitimate in the eternal drama between good and evil. Good, or white, magic was science struggling against the twin enemies of the unknown and pseudoscience, or black magic.

There was, in fact, a good deal of this material sprinkled throughout expert as well as popular literature. In one kind of story, supernatural, unexplained agencies made disembodied use of electrical media, as if to symbolize the mystery of electrical communication itself. *Electrical Review,* for example, quoted the testimony of a train dispatcher fatigued from watching at the deathbed of a coworker, who neglected to direct an eastbound stock car to a siding to avoid a collision with another car carrying a party of excursionists returning from a picnic. As the clock struck twelve, and with seconds to disaster, the dying man appeared to enter the dispatching office, signal each train to a separate track, write down the order in the logbook, and depart. The presumed source of the apparition had expired in his room at exactly that moment, leaving his friend to wonder whether he had written the orders himself while dreaming or been visited by a ghostly presence.[121] Such stories were often presented without comment, simply, as it were, for readers to mull over. Editorial comment framing some stories occasionally even hinted at their validity.

When stories about the occult concerned electricians and their work, however, the editorial presumption was that supernatural explanations were impossible. Perhaps the most characteristic debunking stories were accounts in which exotic, gaudy, spectacular manifestations of spirits appeared before emotionally credulous audiences at scientific demonstrations. Consider this 1897 report from *Popular Science News* of a remarkable Paris performance:

> In a darkened room, a gigantic luminous hand passed over the heads of the spectators, followed by a flock of luminous violins, which we are to suppose were made of glass for the occasion. Then an immense globe descended from the ceiling like a ball of phosphorus, oscillating like the pendulum of a clock. A luminous bell appeared in front of this globe and made regular bows to it. We could see its fiery tongue moving while the globe waltzed around.
>
> Suddenly at the four corners of the room the glasses appeared to become ignited; the vases were illumined and the lustres sparkled. A table loaded with cups and glasses was lighted up. Everything seemed on fire. The whole room, that was so dark an instant before, was aflame on all sides with phosphorescent light of a soft and bluish color.
>
> Then again all was darkness. Gradually a little light streamed in, and in a corner in front of a velvet portiere a human form appeared, at first vague and vaporous, hardly distinguishable. But soon its outlines became clear, and it advanced. The phantom advanced a few steps and then stopped. It was a tall woman. Her face had a greenish pallor; and what an extraordinary face it was. There were no eyes. We could only see two black holes under the eyelids. The mouth was closed, the hair was phosphorescent. A long, luminous veil enveloped this animated statue, and in the folds of the veil little sparks shone like diamonds. She raised her right arm slowly and tossed flames from her hand. A gong sounded. The apparition receded gently, and gradually faded out of sight.[122]

Following this display, a large luminous bouquet appeared in the center of the room next to a blue band upon which the inscription "X-rays" appeared. Lights were turned on. The host of the spectacle rose to explain that nothing in the previous display had an occult or supernatural origin. The entire explanation lay not in sensible appearances but in the knowledge of X rays, "invisible to our eyes." The special properties of X rays caused glass, porcelain, enamel, and diamond objects to shimmer as a beam passed through them in a dark room, while other objects remained concealed in darkness. Phosphorescent sulfate applied to the face, but not the eyes, of a human figure dressed in fluorescent garments glowed in ghostly fashion when X rays were directed at her. The messages of this performance were several. It was clear that science could do magic better than magicians. But the deeper moral was that conclusions reached by untrained lay eyes, especially amid the distraction of mystery, spectacle, and dramatic sound effects, could not be trusted. What could be trusted was detached scientific knowledge originating in considered texts, able to produce not only the effects of the old magic, but other effects that the old magic could never produce.

Late-nineteenth-century electricians constituted a self-conscious class of technical experts seeking public acknowledgment, legitimation, and reward in the pursuit of their task. Their efforts to invent themselves as an elite justified in commanding high social status and power focused on their technological literacy, or special symbolic skills as experts. They distinguished themselves from mechanics and tinkerers, their predecessors, and from an enthusiastic but electrically unlettered public by elevating the theoretical over the practical, the textual over the manual, and science over craft. They sought to define insiders and outsiders in electrical culture, to enforce standards for professional training, and to arbitrate the use of technical language.

As late as 1911 an article in the London journal *Engineering* collected many of these themes into a discussion of the telephone, which was reprinted for a more general audience in *Current Literature*. The article looked forward to the imminent achievement of transcontinental telephony in the United States as a prologue to the day when the spoken word could be transmitted to any point on the globe. That this day would come, explained the editors of *Current Literature*, "is the firm belief of those best qualified to form an opinion as to the possibilities of electrical science."[123] *Engineering* declared:

> The matter is one which must be left entirely to the experts. To the average individual the telephone—like the telegraph, the phonograph, electric light, the steam-engine, and many of the other commonplaces of modern existence—is still a mystery. We avail ourselves of the conveniences and facilities they afford; but how much does the man in the street, to use a convenient term, know of the why and wherefore of the hundred and one scientific miracles which he employs as a matter of course in his daily life.
>
> Take this latest improvement in telephony; what will it convey to the average man to tell him that it has been effected by putting coils of

wire in the circuit of the submarine cable, which have the effect of
setting up in the circuit an inductive action antagonistic to that already
in the cable, or in the circuit; that these two neutralize each other, and
in consequence conversation is rendered possible over a longer distance? . . . It is enough for him to know that the work of the electrical
engineers will enable him to speak with his correspondent at Paris from
Dublin or Glasgow or Edinburgh instead of from London to Liverpool.[124]

Experts also argued that the public, whose desire to share the privileges of electrical elites threatened to dilute their power and prestige,
was misinformed and antitextual. Styling themselves as defenders of
Western civilization, electrical experts caricatured and ridiculed the
aspirations and electrical encounters of conventional scapegoat groups,
including non-Europeans, Indians, blacks, women, criminals, and the
poor. Deception and coercion were accepted sanctions against those
who refused to recognize the authority of electrical expertise. The amount
of deception considered proper was proportionate to the cultural
strangeness of those against whom it was directed. Despite their willingness to exploit public credulity in pursuit of their own goals, electrical experts were deeply suspicious of magicians and performers who
competed with them for public attention and loyalty in the name of
science. Their answer to this challenge was to produce even more impressive magic, some of which is explored in subsequent chapters.

2

Community and Class Order
Progress Close to Home

"Is dis Miss Mandy Johnsing?" asked the voice on the telephone.

"Yas, dis is Miss Johnsing."

"Well, Miss Johnsing, I done called you to de telephone to inquire if you would marry me?"

"Marry you? *Marry* you? Ob course I'll marry you. What made you all think I wouldn't marry you? Ob course I'll marry you. Who is dis talkin', please?"

—Telephony, 1906

Much of the literature on electricity in the late nineteenth century can be read as the wishful template of a world that electrical professionals believed they would create, given the opportunity. The world that looked most comfortable to them was less the egalitarian one they professed to desire in their more self-conscious public moments than one in which the stability of familiar social and class structures would be preserved, with the important exception that other people would come to recognize the wisdom of professional electricians' values, and might even think, look, and act very much like electricians themselves. With the more general application of electricity throughout society, electricians believed, the world could change only to their advantage.

For them, electricity was the transformative agent of social possibility. Through their power over it, it would be a creator of social miracles. Electricity had the vitality of a natural force; they had charge of its control and direction. Experts felt that society had yet to grant them the recognition appropriate to so weighty a social responsibility.

They were also concerned about the possibility of this natural force's getting out of control, particularly their control. Thus, at the same time that electrical professionals were confidently and proudly prophesying utopian accomplishments through the proper exercise of electrical knowledge, especially at the most abstract levels of discussion about the future of civilization they were also conducting an anxious, less publicized discussion about the possible social catastrophes of electrical metamorphosis.

A metamorphosis it surely was, though in the early stages. While electric lights and trolley cars constituted the most visible proof of change in the urban environment, electric telephones had also diffused with impressive rapidity. Early telephone figures are difficult to find and often unreliable. Our best evidence probably reflects only orders of magnitude. Phone figures for the late nineteenth century are rarely broken down into residential and business telephones, or by region, although these were key factors in telephone density and distribution. It is difficult to be sure, after the Bell patents expired in 1893 and the telephone business temporarily opened up to a number of small competitors, how accurate was the reporting of independent telephones. It seems safe to say that by 1899 there were more than a million telephones in the United States. From a base of 3,000 telephones in 1876, or one telephone per 10,000 persons, according to census figures, the number of phones had risen to 54,000 by 1880, and tripled between 1880 and 1884. The number of phones failed quite to double between 1884 and 1893, when the Bell patents expired, by which time there were 266,000 phones, or just under 4 phones per 1,000 population. Passing the million mark in 1899, the number of phones reported in 1900 was 1,356,000, or 17.6 phones per 1,000 population.[1] None of these figures, however, tells us the rate of residential growth, or urban-versus-rural increase.

If this expansion meant progress in the introduction of electricity, it also threatened a delicately balanced order of private secrets and public knowledge, in particular that boundary between what was to be kept privileged and what could be shared between oneself and society, oneself and one's family, parents, servants, spouse, or sweetheart. Electrical communication made families, courtships, class identities, and other arenas of interaction suddenly strange, with consequences that were tirelessly spun out in electrical literature. Because of the official commitment of experts to a future of limitless electrical improvement, whatever ambivalence they expressed did not command the same publicity as their most expansive electrical rhetoric, and was not always acknowledged even by electricians themselves. When speaking of electricity in official capacities and delivering opinions in public forums, electricians were loyal soldiers, laudatory, progressive, and faithful to the cause. In their journals, which monitored the official and unofficial worlds of electricity, and wherever they spoke to one another about what they observed in both realms, electricians gave anxious voice to the possible loss of a world they idealized, a world threatened by new modes of electrical communication and put at risk in the very act of their aspiring to it.

These anxieties appeared mostly as anecdotal, occasional stories of the social order gone awry or threatened with breakdown at vulnerable points, rather than in forthright discussion. According to this textual strategy of indirection, disorder was not structural; it surfaced in small, isolated incidents extracted for their peculiarity from the surface of social life, and appended at the margins of the professional and technical texts that were the central focus of these journals. Nevertheless, the frequency and thematic consistency of these stories were as relentlessly systematic as it was the reassuring function of their anomalous format to deny. Two discussions—one of electrical threat: fragmented, unsystematic, exceptional; and another of electrical promise: systematic, integrated, a sensible model of the very world—were thus carefully segregated from each other. Though occurring side by side, they were framed in different epistemological universes and expressed in different ways, so that their contradictory implications could be conveniently ignored.

Whom Do You Know?

Central to all the good things that new technologies of communication would accomplish was the building of better, usually construed to mean more open and democratically accessible, communities. Listen to *Scientific American*'s glowing prediction in 1880 of the future of the telephone, for which it foresaw

> nothing less than a new organization of society—a state of things in which every individual, however secluded, will have at call every other individual in the community, to the saving of no end of social and business complications, of needless goings to and fro, of disappointments, delays, and a countless host of those great and little evils and annoyances which go so far under present conditions to make life laborious and unsatisfactory.[2]

A decade later Judge Robert S. Taylor of Fort Wayne, Indiana, a prominent American patent attorney and for many years chief counsel for independent telephone interests, told an audience of inventors that the telephone had introduced the "epoch of neighborship without propinquity."[3]

A favorite light sport in popular and professional journals was to detail ways in which timely and imaginative applications of new technologies would unravel social difficulties by opening up new avenues of information. These solutions were presented as though from an observant civic alertness on the part of journal editors, as gratis suggestions to aspiring inventors, and as more or less whimsical fantasies of the future. At the same time, electrical journals chronicled endless complications and disappointments brought about by these technologies, and rendered as new tilts in the world that might signal a social earthquake. Electrical experts worried among themselves, and reprinted the worries of others, that electricity would alter for the worse and forever familiar forms of social knowledge and habits of association, and the usual ways of keeping them under the control of the proper people.

In some cases, technical devices would be able to create communities on the spot. "A phonograph is being made for use at the Exposition of 1900," wrote *Invention* in 1898, "which is expected to be of sufficient dimensions to be heard by 10,000 persons."[4] Using railway figures, *Answers* estimated that in London, with an 1894 population of five million, every twelfth person "in the street, at church, or at the theatre, is a stranger."[5] In technologically modified communities, the stranger was not quite the person he had been before. Who he was remained to be discovered. The city, as Richard Sennett has observed, is a human settlement where strangers are likely to meet, and Michael Schudson has commented that in the nineteenth century, "living became more of a spectacle of watching strangers in the streets, reading about them in the newspapers, dealing with them in shops and factories and offices."[6]

Not even the famous, those who are widely known but personally remote, were exempt from the reorganization of social geography that made socially distant persons seem accessible and familiar. In contrast to *Scientific American*'s utopian yearning for a future community where telephones made everyone available to everyone else was a businessman's account, quoted in *Western Electrician,* of the telephone "maniacs" who plagued the governor of New York, Chauncey Depew: "Every time they see anything about him in the newspapers, they call and tell him what a 'fine letter he wrote' or 'what a lovely speech he made,' or ask if this or that report is true; and all this from people who, if they came to his office, would probably never say more than 'Good morning.'"[7]

The new accessibility fostered by electric media had less than amusing consequences for those who found themselves trapped by the great new machines of publicity. In 1889 the *Evening Sun* related the story of a Cincinnati-to-Washington train ride made on the eve of the John L. Sullivan–Jake Kilrain prizefight by Charles L. Buckingham, "the well-known patent lawyer of the Western Union Telegraph Company." Though the patent lawyer left Cincinnati an "unassuming professional man," by the time he arrived at Washington he had become a famous prizefighter hailed by crowds up and down the line. At Blairsville an eager fan had mistaken him for Sullivan, on account of his tall and muscular bearing, mustache, and "naturally black" eyes, and telegraphed ahead to every station on the line that the great Sullivan was aboard the train. The "unassuming" professional man was surprised, then unnerved, by the crowds that greeted him.

> At every stopping place great crowds had gathered to see Counsellor Buckingham, and even his fellow passengers were certain that he was the great Boston slugger. The latter fact became painfully apparent to the lawyer when, upon walking through one of the cars, he overheard a man remark in a stage whisper to his companion:
> "There he goes now. That man will be as drunk as a boiled owl before night."[8]

In the fantasies of exalted professional men who had achieved their own niche—to say that Buckingham was a "well-known" patent lawyer meant something quite different than to say that John L. Sullivan was a famous prizefighter—perhaps worse than being forced to mix with hoi polloi would be to be mistaken for a drunken Irish brawler, even a famous one.

Home and Family as a Communications Network: Boundaries of Domestic Intimacy

In expert eyes, some of the most radical social transformations appeared to be brewing not around people at a distance, but around those close to home. Particular nervousness attached to protected areas of family life that might be exposed to public scrutiny by electrical communication. That intimate family secrets would be displayed to the world by new instruments of communication was posed as a series of uncomfortable dilemmas. How would family members keep personal information to themselves? How could the family structure remain intact? What could be done about the effects of new forms of publicity on family intimacy? "The home wears a vanishing aspect," lamented *Harper's* in 1893. "Public amusements increase in splendor and frequency, but private joys grow rare and difficult, and even the capacity for them seems to be withering."[9]

Popular Science News offered an example of that nightmare come true in 1901. In Detroit, moving-picture footage of the occupation of Peking showed a detachment of the Fourteenth U.S. Infantry entering the gates of the Chinese capital.

> As the last file of soldiers seemed literally stepping out of the frame on to the stage, there arose a scream from a woman who sat in front.
> "My God!" she cried hysterically, "there is my dead brother Allen marching with the soldiers."[10]

In the romantic fashion in which absent male protectors were often explained away, Allen McCaskill had mysteriously disappeared some years before. The War Department soon confirmed that the man "whose presentment she so strangely had been confronted with" was indeed her brother. The fascination of the story was not only its invitation to the reader to imagine what it would feel like to receive so grave a shock so unexpectedly, or to ponder the ease with which the distant could be made proximate, but also its vivid account of private emotion displayed before strangers in a public setting intended for vicarious entertainment.

Feared above all for its potential to expose private family secrets was the telephone. In 1877 the *New York Times* discussed its "atrocious nature" in an account of how experimenters had stretched telephone wires across four miles of Providence, Rhode Island. With uneasy humor, it reported that the telephone men heard eloquent clergymen, melodious songs, midnight cats, and "other things which they did not venture to openly repeat," including "the confidential conversations of hundreds of husbands and wives."[11] It would be a "hazardous matter" for any resident of the city to accept nomination for office in light of these possibilities, the *Times* concluded. "We shall soon be nothing but transparent heaps of jelly to each other," a London writer fretted in 1897, after two more decades of telephone development.[12] Boundaries marking public and private seemed to be in peril as never before.

Elsewhere such presentiments were reality. Upon discovering that his residential telephone was engaged whenever he called home, a suspicious husband had the phone tapped and discovered that his wife was having a love affair. The telephone conversations between Mrs. Trowbridge and her paramour, Jonathan Ingersoll, were accepted as grounds for divorce in a New Haven court in 1889. Following this case, the Connecticut Telephone Company asked for, and the Connecticut legislature passed, a law against wiretapping that imposed a five-hundred-dollar fine on violators. Two New York telegraph operators accused of tapping Western Union wires for racing information to secure a betting advantage over a New York turf commission house were brought to trial in the first case prosecuted under this law, in 1891.[13] "The courts of justice cannot and do not desire to ignore the great changes . . . which the introduction of the telegraph and telephone have accomplished," declared Judge Shepard Barclay in *Globe Printing Company* v. *Stahl,* an 1886 case in Missouri involving the admissibility of telephone conversations as evidence.[14] Telegraph companies had long taken the position that their dispatches were "privileged, like the confessions of a patient to his doctor or a parishioner to his priest," or like the mails, and therefore not to be produced in court, or for any public consumption. The *Electrical Review* speculated on the dangers of abandoning this model of privileged communication:

> Would any rival have a right to tap the wires between two lovers in order to satisfy himself that the "other fellow," and not himself, had been the fortunate wooer? Suppose one business concern desired to learn the secrets of a competitor. Would it have the right to tap the wires and take advantage of the information thus acquired?[15]

Not everyone was so pessimistic. A standard line of argument held that divorce would decline with the resurrection of home life made possible by the electrical decentralization of industry and the electrical mechanization of domestic chores such as cooking. "'Talk about your Roentgen rays,' said one woman recently," speaking for this segment of opinion in a popular women's magazine, "'If people would pay half as much attention to this cooking feature of electricity there wouldn't be so much talk about marriage being a failure.'"[16]

Nevertheless, the picture that emerges from less self-conscious accounts in the professional literature is one of the bourgeois family under attack. New forms of communication put communities like the family under stress by making contacts between its members and outsiders difficult to supervise. They permitted the circulation of intimate secrets and fostered irregular association with little chance of community intervention. This meant that essential markers of social distance were in danger, and that critical class distinctions could become unenforceable unless new markers of privacy and publicity could be established. Energetic efforts were made, therefore, to limit opportunities for new instruments of communication to create new secrets, and to protect the old ones they put at risk.

Irregular Courtship and Lightning-Struck Romance

"The invention of new machinery, devices, and processes," reflected *Telephony* in 1905, "is continually bringing up new questions of law, puzzling judges, lawyers, and laymen . . . . The doors may be barred and a rejected suitor kept out, but how is the telephone to be guarded?"[17] How indeed? New forms of communication created unprecedented opportunities not only for courting and infidelity, but for romancing unacceptable persons outside one's own class, and even one's own race, in circumstances that went unobserved by the regular community. The potential for illicit sexual behavior had obvious and disquieting power to undermine accustomed centers of moral authority and social order.

Other changes in the social conduct of love were not so ominous. Electrical professionals were often amused by electrical reconfigurations of romantic courtship, defined by them as the first exploratory step in the orderly initiation of new families. They played lightly on the theme of romantic awkwardness made more excruciating or magically banished by electric communication. Typical stories related young love betrayed by unaccustomed technology. Sometimes poor transmission made impassioned lovers' speeches seem to issue from inebriated lips, as in the case of a student of elocution rejected by the lady of his dreams, who had heard only slurred syllables over the telephone instead of the poetic recitation her suitor intended, and "now declines to have anything to do" with the unfortunate young man.[18]

Another tale of frustrated courtship was set in Atlantic Highlands, New Jersey, a small resort village of "500 souls." During the summer of 1886 a "nice young man" from the city met "one of the rustic beauties of the place" and fell in love. They corresponded, and she invited him to visit. One day a telegram appeared with news of his impending arrival.

> Somehow—nobody ever will know just how—fifteen minutes after the
> message clicked into the office every person in town knew that young

Blake was coming to see Miss Trevette. Every young lady of the town made up her mind to catch a glimpse of this rash young man who sent telegrams, and every man determined to be there to see that everything went smoothly.[19]

When young Blake alighted from his carriage, full of the foreboding that marked him as a proper suitor, an audience of 499 villagers had gathered to watch. They observed while he paid the driver, studied him as he asked directions to the young lady's house, and followed his progress up the hill. Panicked by the approaching procession, Miss Trevette sent word of her absence, halting the romance at a blow.

Observing every propriety, this couple had kindled their romance in person and developed it in a decorous written correspondence. Electric communication had set up the next stage of romantic involvement but, despite its glamor, proved too clumsy for the subtleties of courtship. Once launched, this story suggested, electric romance could not return to a slower and more innocent state. This tale also celebrated the triumph of tradition over youthful experiments with cherished community customs. The citizens of Atlantic Highlands were attached to the young lady in question by older and sturdier bonds than was her distant suitor, and were connected to one another by a practiced network of communication more efficacious than any technological marvel.

Traditional courtship protected traditional young women from inappropriate advances by placing insurmountable obstacles in the path of all but the most devoted swains. Electrical communication threatened to shortcut the useful insulation of these customary barriers. Thus, the expert press viewed with uneasiness and fascination the new breed of women who were the cheerful beneficiaries of new possibilities for contact between the sexes. These women stood in sharp contrast to the Miss Trevettes of the world, who were unwilling to embrace changes in the rules too quickly.

The new woman's wider contact with the world made her brasher and more resilient. Her prototype was the telephone girl, the operator who dealt daily with the demands of male strangers. If she were an "all-night" telephone girl, her friends might be the inventive young college men who were night clerks in the "always-open drug stores."[20] Moving in a larger community where personal acquaintance did not provide the usual safeguards, the telephone girls improvised ways to protect themselves from unwanted attentions of a kind they might not have otherwise encountered. They assumed, for example, the same privileges as men in sharing compromising information about the liberties of would-be suitors. The telephone girl was "not afraid to confidentially give you some points concerning Brown, the young broker, who is considered a lady killer of the first water. Brown has bored several of the girls to death with his attentions."[21]

Not only telephone girls were tempted by new media to flout the conventions of obedient behavior generally expected of women. The "matinee girl," a *fin de siècle* phenomenon, also symbolized the risks of letting young women loose in an artificially enlarged world of seductive messages for which their training and background had not prepared them. It was feared that their heads might be turned by the wrong sorts of men, men beyond the social boundaries of the community, men with superficial appeal but no substantial roots.

> Only in recent years has the privilege of attending theaters been allowed to young women unescorted by brother or father, or the creature known by the elastic and accommodating term "cousin." This last decade of the nineteenth century has brought to girlhood the inestimable joy of going wherever her own sweet will dictates. . . .
>
> But undoubtedly the greatest and most important of all reasons for the girl's fondness for the matinee is the fact that it affords the impressionable maiden from two to three hours of almost uninterrupted adoration of her hero, the leading man, the stage representation of all that is correct, all that is good form, all that is perfect. . . .
>
> The matinee girl is a creature of whims. Her very existence is nothing but a fad. Her heroes are only fads, and fleeting ones at that, for they change with every season, or, oftener, with every metropolitan success. . . . Her elastic heart is ever ready to shape itself to another ideal.[22]

In contrast to the challenge new channels of courtship offered to community custodianship of romantic custom were stories of romantic competition, where seizing a technological advantage in the fierce rivalry for female affections was a sign of masculine prowess. *Harper's Bazar* related an encounter between two telegraphers and a newlywed couple traveling on the same train. By tapping their pocket knives in Morse code on the metal armrests of their seats, the telegraphers made ribald comments about the charms of the bride and the lack of them in the "chump" she had married. Escalating their outrageousness, one proposed to kiss the bride when the train reached the next tunnel, cuckolding her new husband in the way of a practical joke.

Their telegraphic conversation ceased here, for the bridegroom had taken out his pocket knife and commenced to tick off this passage on the arm of his seat:

"When the train gets to the next tunnel, the chump proposes to reach over and hammer your heads together till your teeth drop out. See!"[23]

A frequent anecdotal character was the "enterprising" suitor who ran a modern courtship.[24] The enterprising suitor offered his proposal of marriage by telephone at the same moment that his unsuspecting and less courageous rival waited anxiously in the parlor, only to discover, upon his hostess's return, that her hand was taken. The electrical journals shed few tears for the old-fashioned suitor who relied on love's gradual unfolding. Since the women in these stories seemed ready to accept the first proposal they received, the race was to him who pressed his electrical advantage. An obviously apocryphal but typical account from the *New York Sun* went as follows:

The young man had been trying to tell her how madly he loved her for over an hour but couldn't pluck up the courage.

"Excuse me for a moment, Mr. Featherly," she said, "I think I hear a ring at the phone." And, in her queenly way, she swept into an adjoining room.

Presently she returned and his mad passion found a voice.

"I am sorry, Mr. Featherly," she said, "to cause you pain, but I am already engaged. Mr. Sampson, learning that you were here, has urged his suit through the telephone."[25]

Irregular courtship threatened the orderly exit of children from the family into approved relationships with the opposite sex. As often, however, as new media presented opportunities to stray from the family circle without detection, they provided more frequent occasions for parental surveillance of actual courting behavior. One suspicious patriarch placed a phonograph under the sofa and recorded an intimate exchange between his daughter and her new beau, which he played to the family at breakfast, especially this choice excerpt:

"Are we alone, dear?" . . .

"Yes, Cholly, but now don't." . . .

"Oh, come on, now; don't be skittish," and Sis's voice says, "Now Cholly, quit. Someone'll hear you." Then they kept on.

"You shan't kiss me."

"Oh, yes, I shall."

"No, you won't."

"Well, then, there," and a kind of screechy noise, like a cab driver starting his horses, came out.[26]

In other stories, the escape from parental supervision made possible by new communications technologies carried great risks. This was the bitter lesson learned by George W. McCutcheon of Brooklyn, the father of twenty-year-old Maggie, who found himself pleading not guilty in 1886 to "a charge of threatening to blow her brains out." When McCutcheon's newsstand business prospered and he decided to expand his operation, he set up a local telegraph station in a corner of his store, with his daughter as its operator. He soon discovered that Maggie was "keeping up a flirtation" with a number of men on the wire, including one Frank Frisbie, an operator in the telegraph office of the Long Island Railroad. Maggie issued Frisbie a telegraphic invitation to call on her at home, which Frisbie accepted, also by wire. Maggie's father forbade the visit, and Maggie began to see Frisbie, a married man with a family in Pennsylvania, on the sly. Hoping to thwart his daughter's illicit suitor, Maggie's father moved his store to a different location. This sacrifice was futile. With her expert skills, Maggie easily found work at another telegraph station and resumed the relationship. Her frustrated father pursued her to another rendezvous and threatened her with bodily harm. She had him arrested on the spot.[27]

Here was a story to chill the parental heart. It exposed once more the perils of training young women in the skills of new communications technologies. Maggie had flirted with strange men under her father's very nose. Frank Frisbie was a railroad telegrapher of more than passing acquaintance with another community-shattering technology of the nineteenth century, transportation, which perhaps had facilitated his desertion of his own family. Telegraphy provided similar protection from conventional parental and community tests for dubious romantic involvement. In removing daughter and livelihood to a different location, Maggie's father acted like any respectable paterfamilias who feared for his daughter's reputation and future happiness. Equipped with the skills of independence, Maggie could outmaneuver her father and go where she pleased, however poor her judgment. In the end, the disruption of the family was complete. Father did not bring daughter to heel; daughter brought father to the authorities in full and disgraceful public view.

Another story of cross-class romance described how an "ardent Gentile suitor" wooed "a beauteous, blushing Jewess" against the wishes of the girl's family. Forbidden to meet or correspond, the couple eloped after "spending hours and hours every day cooing to each other over a telephone line between the offices in which they worked, and by that means they had planned all in secret."[28] *Electrical World* pondered the future of star-crossed love deprived of romantic obstacles:

> The serenading troubadour can now thrum his throbbing guitar before the transmitter undisturbed by apprehensions of shot guns and bull dogs. Romeo need no longer catch cold waiting at Juliet's balcony, nor need Leander encounter the perils of the waves in seeking his Hero. But the new means of communication has its disadvantages, and must be used with caution. There will be the danger of a wrong "connection" at the Exchange, and there will be no letters to produce as evidence in a breach-of-promise action.[29]

Letters there might not be, but alternatives could be improvised in telephone taps and phonograph recordings. Romantic overtures made in secret and, as an impetuous young man might imagine, without witnesses were no match for the phonograph. Consider a story of the precautionary use of technology against irregular courtship practice, in this instance the seduction of the vulnerable by the cunning. The solution worked to the advantage of the truly cunning—the class of literate, expert professionals.

> "Did I ever say all that?" he asked despondently as she replaced the phonograph on the corner of the mantle-piece. "You did." "And you can grind it out of that machine whenever you choose?" "Certainly." "And your father is a lawyer?" "Yes." "Mabel, when can I place the ring on your finger and call you my wife?"[30]

Comparing this account of phonographic enforcement of proper courtship practice to the story of "Cholly" and "Sis" also tells us something about perceived class differences in spoken and literate modes. The cues of colloquial speech label Sis's family as lower class, and Sis's father reasserts the limits of proper courting behavior through oral humiliation. The lawyer's daughter, on the other hand, is protected by her father's knowledge that complex legal procedures can bring an errant lover to heel. In one family, authority is spoken and personal, and the threat of exposure is limited to the family circle. The other family looks to the law, a literate, impersonal point of reference, and exposure is threatened before a professionally literate community of strangers who, learning of a young suitor's misbehavior, could ruin him. For one couple, the issue is whether the behavior in question departs from proper custom. For the other, it is whether the behavior has the status of a written contract.

Protecting the Domestic Hearth

Since electricity in the home was situated at the family's private and vulnerable center, it was necessary to domesticate it. The electric light, for example, was praised for its positive, middle-class virtues of "beauty, purity, brightness, cleanliness, and safety" compared to other forms of domestic illumination.[31] "Gas not only consumes and pollutes the air," the *Electrical Review* explained, "but is very poisonous, besides having a deleterious effect on the furniture and decorations of our homes." Gas fumes blackened ceilings and cornices, and discolored wallpaper. Electricity created a brighter, cozier, safer, more intimate haven.

Electrical journals argued that electricity would replace household servants by mechanizing domestic tasks, and touted it as the "key to the servant difficulty." Electricity would free the mistress of the house from time-consuming chores and help settle once and for all the question of "woman's sphere"—though woman's sphere was defined proximate to electricity with great reluctance by male experts.[32] The residential telephone was rarely included in the litany of electrical devices that would replace servants, however. Instead, the telephone must have increased servant's responsibilities for filtering communication between outsiders and householders, since servants often were buffers between importunate telephone callers and household members. Used mainly to talk with one's own social set, residential telephones did not perform any job normally handled by servants. Outside the home, electrical communication might revolutionize the world; inside, it should make the old world work better and as it had always been intended to work—on a successful class model. Inside the home, the chief purpose of electrical communication was to improve the facility with which orders were given and received. Innovations in domestic communication were intended more closely to bind servants to masters, wives to husbands, and children to parents.

Home was the protected place, carefully shielded from the world and its dangerous influences. New communications technologies were suspect precisely to the extent that they lessened the family's control over what was admitted within its walls. Householders resisted both the symbolism of outside intrusion and its physical expression in wiring, a tangible violation of intact domesticity. *Fin de siècle* comparisons of electrical progress between domestic and occupational settings remarked on the reluctance with which homeowners gave permission "to tear up the house to have the wires strung."[33]

Electric burglar alarms were regarded as an appealing domestic innovation, but any hint of technical complexity raised doubt and suspicion. "To surround the Englishman's castle with a network of electrical wires is as little to his taste as would be the arming of it with a battery of four-pounders," wrote the *Electrician*.[34] This was partly because it was not clear that he could understand them. Unlike the stigmatized classes whose ignorance of electricity made it easier to control them, the classes whose well-being new electrical devices were supposed to serve wanted them to be simple and reliable. It would not do for respectable householders to feel victimized by a technology they had adopted to increase their mastery of the world. Electrical devices must fit unobtrusively into the household routine. Alarm batteries must not be messy, and it must be possible to conceal them so that "the most fastidious of men may be satisfied." Nothing of domestic life should warrant drastic adjustment as a result of electrifying it.

Discussions of electricity as a staple of domestic routine emphasized the economically and morally self-sufficient household, a productive unit beholden to neither neighbors nor bosses, industriously and privately going about its own business. This was the image reflected in an anticipation of the future written by Henry Flad in the *St. Louis Globe-Democrat* in 1888:

> The time is not far distant when we will have wagons driving around with casks and jars of stored electricity, just as we have milk and bread wagons at present. The house of the future will be constructed with the view of containing electric apparatus for lighting, power, and cooking purposes. The arrangements will be of such a character that houses can be supplied with enough stored electricity to last twenty-four hours. All that the man with the cask will have to do will be to drive up to the back door, detach the cask left the day before, replace it with a new one, and then go to the next house and do likewise. This very thing will soon be taking place in St. Louis.[35]

The author of this image had attached his intuition of a new order to the landmarks of a thoroughly familiar and archaic one, in this case door-to-door distribution by hand and horsepower of a commodity as perishable as milk or ice. This image of the inviolable container of the family was consistent with the grandest visions of the properly appointed home of the future.

Perhaps its most lavish expression was the Greenwich, Connecticut, home of "electrical millionaire" E. H. Johnson, a successful Edison partner, whose country mansion on three acres was what every self-sufficient electric house aspired to be. Its most dramatic expression was visual; at night it fairly glowed with electric light. The machines that supplied current to its many decorative and functional lamps also registered the temperature indoors and out, the force of the wind, the force of the electric current, and the pressure on the boilers. Electricity blew fresh air into the house, controlled its atmosphere with dampers and ventilators, and, of course, protected it from intruders. "Let any one try to open the gate to the park or seek an entrance to the mansion by door or window, and he will set the secret forces to blabbing loudly of his folly." Soon electricity would "work the churns, make the ice for the household, drive the lawn-mower, and do a score of helpful things about the house and grounds."[36]

In 1893 *Answers* asked its readers to describe their ideal homes. Featured in several issues under the title "Which One of These Suit You? Ideal Houses with More Than the Latest Improvements" were eleven entries. Seven mentioned electric appliances, fittings, or communications devices. Only one described a system of electric communication with the outside world.[37] Typical was a romantic fantasy of a feudal fortress against the world, where aristocratic inhabitants effortlessly communicated with one another but not at all with outsiders:

Something Like a Castle

In outward appearance something like a castle, surmounted by a tower, walls 18 inches thick and covered with ivy and Virginia creeper, French-windows and plenty of them, doors and window frames of mahogany, floors oak.

On the ground floor, ample hall with Roman archways, fire appliances handy; library and drawing-room Early English style, the latter with small nooks to seat two or three, hung at the entrances with curtains, and lighted at night with fairy lamps; dining-room and domestic offices, modern, smoke-room Oriental, with settees round walls and small tables for coffee; bijou theatre, Indian, billiard-room, gymnasium, laboratory.

Upstairs (one floor only), bedrooms and dressing-rooms, bathroom with plunge bath. In tower, barroom and nursery, observatory at top.

Balcony around the building level with the upper floor, with five ladders ready for lowering; speaking-tubes with electric bells all over the house; hot-air warming apparatus throughout, also open grates.

Electric light with slightly tinted globes; hot and cold water laid on to bed and dressing rooms, stabling, lawn, kitchen and flower garden, conservatory.

The sole fantasy house with communications links to the outside was described as "What a Lady Wants." Electric bells "throughout" provided internal communication. From the lady's boudoir, telegraph wires were connected to the railroad station, and a telephone was connected to her husband's office. Electric pushbuttons, phonographs, and telephones were all present in the most technologically elaborate house of all. Predictably, all telephones were for internal communication; pushbuttons and phonographs mediated the only outside contact.

The Empty House of 1993

To let, a most desirable residence, standing in its own grounds. Large lake for rowing and swimming; artificial ice on skating rink.

The house fitted throughout with electricity, electric stoves in every room, improved electric cooking range in kitchen.

All the stoves can be lighted by pressing a button at the bed-side.

Doors and windows fitted with electric fastenings, phonographs for communicated messages fixed to front and back doors.

Every room connected by telephone. Latest fire-escape appliance, namely, side of house swings out by electric power, and large water-cistern, with fire-sprinklers, covers the whole roof.

Grand promenade on top, reached by lifts both inside and out. For particulars, apply—

In this house callers' messages were stored phonographically on an aural calling card. Whereas telephone subscribers entered into a continuous relationship with the outside world by their implicit agreement to endure inconvenient and unceremonious interruptions, phonographic messages were received entirely at the pleasure of the phonograph owner. The decorum of domestic sovereignty was preserved. The phonograph was frequently praised for its contributions to a middle-class ideal of family life in a way that the less governable telephone was not.

Nothing is more wonderful, nothing more fascinating than the exploits of a well-trained phonograph. Faithfully melodious, it reproduces again and again the strains sung or prose tit-bits recited in domestic circles; and this sort of thing gives the phonograph a sentimental value that it is difficult to appraise. Appealing thus, to the deepest laid instincts in our nature, there can be no doubt that the phonograph—well made, and put on the market at a price which is reasonable when considered in respect to the mechanical nature and detail of the instrument—will come

into world-wide demand. No well-kept, intelligently cared for home will
be considered complete without its phonograph.[38]

The desire for some tangible imprint of domesticity on media was
metaphorically rendered in a speech by Emile Berliner to the American
Institute of Electrical Engineers in 1890. He looked forward to the day
when sound recordings would be made from etched steel matrices, and
possibly cast in glass: "We may then have dinner sets, the dessert
plates of which have gramophone records pressed in them and which
will furnish the after-dinner entertainment when the repast is over.
Gramophone plaques in the voices of eminent people will adorn our
parlors and libraries."[39] Media were to adorn, not transform home life.
Home was still the quiet place of culture where the routines of women
would go on supporting men's forays to less protected places.

This social logic was suspicious of "aesthetic" contributions by
the telephone to life around the domestic hearth, which Edward J. Hall,
the new vice-president of the fledgling American Telephone and Telegraph Company, hinted at in 1890:

> More wonderful still is a scheme which we now have on foot, which
> looks to providing music on tap at certain times every day, especially
> at meal times. The scheme is to have a fine band perform the choicest
> music, gather up the sound waves, and distribute them to any number
> of subscribers. Thus a family, club, or hotel may be regaled with the
> choicest airs from their favorite operas while enjoying the evening meal,
> and the effect will be as real and enjoyable as though the performers
> were actually present in the apartment. We have perfected the distri-
> bution, and have over a hundred . . . persons who have certified to
> their anxiety to be subscribers.[40]

Hall admitted that a number of technical problems needed to be solved,
since the telephone could not successfully distinguish among harp, piano,
reed, wood, or brass tones. Some in his audience feared the success
of this scheme far more than any imperfection of transmission technology. *Electrical World* objected:

> Fancy turning on the music at will and listening for an hour to the splen-
> did performance of a famous orchestra. The idea is most luxurious and
> attractive, fit ornament of a symbolic age. And yet if we look further,
> beyond the mere outline of the suggestion there lies before us a vista
> of dreadful possibilities. For with the success of the first telephone mu-
> sicale association there will spring into being rival organizations, the
> very names of which would make incipient deafness bliss. Imagine the
> awful devastation that could be wrought by "The Organ Grinders' Tel-

ephonic Mutual," with a drop-in-a-nickel slot attachment. Fancy the horrors of having one's disposition wrecked by a "popular programme," headed by a memorial to the late McGinty. And what *new* terrors would be added to that Juggernaut of the metropolis, the boarding-house, when "Sweet Violets" and other appetite-destroying tunes could be turned loose at feeding-time. . . . The probability of cut-rate competition and an orchestrion in every boarding-house is really too horrible for contemplation.[41]

What was exclusive and luxurious was domestically desirable. What was popular and Irish was not, and what was at stake was domestic peace at mealtime, one of the most stridently defended battlefields in the struggle of the middle-class family not to fragment into the rootless chaos of the boardinghouse.

Other threats to the middle-class hearth were still more palpable. The *Electrical Review* related the story of a "wealthy, well-educated, and fashionable [Chicago] woman who resides in a handsome home on one of the aristocratic South Side Avenues," in search of a house-keeper to care for her children so she could leave town to take care of a family emergency. A housekeeper in service to another family recuperating from scarlet fever was recommended to her, and she was urged to expedite arrangements by telephone. At first she was "aghast at the proposition, and was sure that there would be great danger of infection" by wire, her fears a metaphor for all the elements of the world beyond domestic control. After weighing the arguments of a knowledgeable friend, she concluded: "Well, I suppose I must risk it. I'll have a servant call up the house and tell them be sure that the housekeeper changes her clothes and the sick children aren't in the room where the telephone is; then I may feel justified under the circumstances in talking with her."[42] As late as 1894, *Electrical World* reported that the editor of a prominent Philadelphia daily newspaper had cautioned his readers not to converse by phone with ill persons for fear of contracting contagious diseases.[43]

Communication Between Masters and Servants

Outside the home, electrical communication was the province mostly of men, into which territory the occasional misguided or foolish woman strayed, and from which she had to be protected and now and then rescued. Within the home, the emphasis was on enhancing domestic tranquillity and improving the facility with which messages passed from masters to servants. Troubled by his maidservants' inattention to the bell that summoned them, a reader of *Science Siftings,* a weekly British periodical for lay readers interested in science, applied a "scientific turn of mind" to the problem for a "Queries and Answers" column that awarded prizes for submissions from readers. He proposed to put "Hertz-wave emitters" in his sitting room and bedroom. Each maid carried a small battery, a tiny electric bell, and a Hertz-wave "detector" in her pocket.

> Now, when I require my morning coffee, hot water, or what not other little convenience, I must but touch a button, sparking my bedroom wave emitter once, so that I may be almost instantly attended to by one or all of my domestics, from wherever occupied in or about my premises. . . . Or if I were in my study and required any service, all that I should have to do would be to press the button, we will say twice, and have conferred upon me quite a little stream of blithe assistance from all quarters in my house. These are my views, as to what it may be possible for the luxuriously situated bachelor to accomplish.[44]

An innovation in the household of a wealthy ironmaster in northern England, "whose house and works are dazzlingly illuminated by the electric light," was reported in the British *Court Journal* as a "grand invention" to keep track of servants. This consisted of cameras concealed throughout the rooms of the house, their purpose to "glean some information as to what goes on during his not infrequent absences from home." Each hour shutters were silently opened by machinery and a photograph was taken of everything in the room. On each return home, the master enjoyed developing these pictures, to the occasional dismay of the servants:

> One clerk, who received his dismissal somewhat unexpectedly, and boldly wanted to know the reason why, was horrified when shown a photograph in which he was depicted lolling in an easy chair, with his feet upon the office desk, while the clock on the mantelpiece pointed to an hour at which he ought to have been at his busiest. The servants' party in the best dining room furnished another thrilling scene.[45]

A more widely known innovation was the "talking clock," or "phono-clock," a phonograph recording triggered by clockwork, regarded as especially useful not for replacing the servants one had, but for monitoring them more successfully. It was said that the inventor of the talking clock "especially aims at getting lazy servants out of bed, and has constructed the clock's vocal apparatus so that the purchaser's voice can be imitated."[46] The task of the talking clock was to rouse sleepers and get them to work without exception or forgiveness. Almost every account stressed the indignity and relentlessness of its ministrations.[47] If recalcitrant victims did not respond to endless shouted commands, these could be supplemented by special electric gimmicks. For example:

> At the recent Leipsic [sic] Fair an interesting early-rising appliance was exhibited. By means of a strong electric current the occupant of the bed is twice aroused by the ringing of the bell, after which a tablet with the words, "Time to get up!" is thrust before the sleeper's eyes, then his nightcap is pulled off his head, and last of all—if not thoroughly awakened by this time—he is pitched mechanically out of bed on to the floor.[48]

Much of the amusement occasioned by the clock lay in its target's inability to dawdle, object, bargain, sulk, or argue with effect, that is, to negotiate with the perpetrator of his or her misery. In a certain sense, the phonographic clock was simply a domestic variation on other well-established strategies devised by middle-class householders to keep outsiders from addressing them. Its appeal derived from the ability not to render communication more effortless but to render others speechless and make discourse as difficult as possible. The range of applications for making servants, children, and spouses powerless to answer back was very wide. On this topic *Answers* speculated:

> In the nursery the solemn timepiece could be made to say, "Children, it is time to get up; dress quickly, and do not dawdle." In the kitchen it would be ready with, "Breakfast at eight sharp, Mary; don't forget," and in the dining-room, "You must start in ten minutes or you will lose your train." . . . The dial of this clock of the future is, we are told, a human face, from whose uncanny mouth comes the announcement of the hours, as well as any directions which may be left with it.[49]

The serenity that the bourgeois household prized and that gave rise to strategies for sequestering the family from intrusive media was not a priority in managing the servants, who were the most acceptable targets of the raucous talking clock. Occasional efforts to move the talking clock up the social scale, therefore, changed its characteristics accordingly. "What is the most up-to-date clock?" asked *Science Siftings* in 1899. "The timekeeping phonograph retains easily first place among clock curios of the day. This phono-clock announces the time, in a restrained yet clear voice, every five minutes, so that it proves one of the most soothing ameliorations of the bedchamber."[50]

The unaccommodating phono-clock contrasted sharply with the practice in some cities of having telephone operators awaken male subscribers. In an age when female telephone operators were viewed as a kind of personal servant to subscribers, the "hello girls" often acted as personal alarm clocks. While this service was presumably available to mistresses as well as masters in subscribing households, most stories depicted friendly, bordering on suggestive, relations exclusively between the telephone girl and her male customers. One exchange went as follows:

> "Hello, girlie," he gurgles to the sweet voiced operator at the other end. "I want to get up at 6:30 to-morrow morning. Will you be so good as you sound and ring me up then? If so, there will be something in your stocking about Christmas time. Ever go to the theater?"
>
> That's the way it begins. The telephone girls are an accommodating lot, but even if they were not there would be trouble if they failed to awake several thousand Bostonians every morning of the summer, for here it is a rule of the company that they accommodate patrons.[51]

The accommodating telephone girls were outsiders brought inside on a model of domestic servitude, that is, under the potential control of the rules of the bourgeois household. "The telephone girls may fairly boast of being connected with the best people of the city—by wire," punned the *Boston Transcript* in 1888, with a finger on the telephone girl's tricky and worrisome social status.[52] It was necessary to present her as a socially competent performer, a smooth and knowledgeable broker of social relations between middle-class households, and to make clear at the same time that she was only a servant, not truly a member of the class to whose secrets she had access.

Offering domestic servitude as a model for interpreting the telephone girl did not quite create the reality, however. Telephone girls were definitely not within the household, but quite outside its rules in important ways. Unlike true domestic servants, they served more than one household. As young women earning a living independently of any household they served, they were not subject to the taboos or claims of loyalty binding those within its walls. They were often objects of fantasy, providing "little glimpses of life at other seasons forbidden."[53] "One of the young ladies at the Central office has a singularly pleasing voice, and it is just possible that her features match it," went a typical whimsy.[54] The predictable problems arose. Jones, a Boston businessman, had solicited the services of the "sweet-voiced" operator while his wife was on summer vacation. Mrs. Jones returned suddenly, and had a surprise waiting.

> The next morning the telephone rings. Horrified hubby sits up in bed in dazed surprise, while Mrs. Jones goes to the 'phone.
> "Hello," says Mrs. Jones.
> "Hello, pet," comes back in a woman's voice. "Hurry up and get up. I've been ringing for you long enough."[55]

The triangle of husband, wife, and telephone girl, in which one or more of the parties was confused about the role of the telephone girl in the husband's life, was what most of these stories were about. Usually they revolved around apparently innocent remarks made by the husband and misinterpreted (but not entirely) by the wife. In such stories, sexual decorum was usually restored by the wife, who asserted her distaff prerogative firmly. A story was told of "a gentleman who holds public office and . . . was talked of for mayor." His clerk always spoke to Central, and never

> without a term of endearment. The discussion over the wires generally began with: "Is that you, dear?" and wound up with "Good-bye, darling!" In the absence of the clerk, the distinguished public man went to the telephone in person. Central promptly answered, and failing to recognize the voice asked, "Is that you, dear?" "No, darling," responded the distinguished public man, "it's the other fellow." It is the good fortune of some wives to make their appearance just in the nick of the most exasperating time. This is what happened in this instance. Behind the distinguished public man, when he said, "No, darling, it's the other fellow," stood his wife, who had concluded to visit her husband that morning. She startled him by exclaiming, "Well, I like that!" Did she?[56]

Tension and Trust: Asymmetry and Class Consciousness

Fears of new forms of communication were fears that exclusive communities would be overwhelmed by outsiders against whom there could be no defense. Class and family boundaries could be maintained only so long as inadmissible outsiders were effectively screened. Some stories in the electrical literature depicted outsiders as predators with evil, even criminal intent toward community members. Others portrayed them as masqueraders lowering the tone of insider communities, and still others as inconsequential bumblers, easily dismissed. Asymmetries of dress, manner, and class that identified outsiders and were immediately obvious in face-to-face exchange were disturbingly invisible by telephone and telegraph, and therefore problematic and dangerous. Reliable cues for anchoring others to a social framework where familiar rules of transaction were organized around the relative status of the participants were subject to the tricks of concealment that new media made possible. Lower classes could crash barriers otherwise closed to them, and privileged classes could go slumming unobserved.

Even when no deliberate effort was made to disguise class differences between two parties connected by telephone, that connection was still an unprecedented breach of the normal social insulation that protected higher classes from association with lower ones. These new social facts paved the way for debacles real and imagined. In 1889, a man entered the long-distance telephone office in Syracuse, New York, to announce that he wished to speak to Mrs. Cornelius Vanderbilt, Jr., and Mrs. Levi P. Morton and her daughters. He claimed to be on intimate terms with all of New York society's "400" as well, saying he spoke with them frequently by telephone. He was locked up as demented, having been too insistent in his fantasies of familiarity with those who were socially taboo.[57]

Royalty, whose members stood behind the most impenetrable of class barriers, provided the most dramatic backdrop for stories about movement by members of different classes into and out of one another's territory with the aid of electricity. When telephone wires were laid from Buckingham Palace to London theatres and concert halls in 1896, the *Electrical Review* was moved to comment, "It will even be possible for the royal ears to hear the latest music-hall gags, and that, too, without compromising in the slightest degree queenly dignity."[58] More intriguing than the prospect of a queen surreptitiously sampling the popular culture of her subjects were stories in which royal barriers were trespassed by technological interlopers, particularly brash Americans with no tradition of class deference. In 1905, an official diplomatic letter of introduction failed to get a party of visiting Americans past the king's stables and into Buckingham Palace without the express consent of the absent lord chamberlain. The undaunted Americans insisted on having the palace telephone connected directly to King Edward, who was attending the races at Newmarket. This request astonished the functionaries at each of the well-padded levels of bureaucratic protocol designed to protect the king, but its very exceptionalness left them without procedures for denying it. In a very short time, "by his majesty's commands" the Americans were granted permission to "do" the palace, attended with all courtesies implied by the king's personal permission.[59]

Proprieties of Presence

Doubts about the motive and station of the person with whom one was engaged over the telephone often focused on how to interpret remote or nonimmediate presence, that special form of interpersonal engagement peculiar to new media. Behaving as if new forms of presence were just the same as those they purported to extend invited serious confusion, since the social clues supplied by remote presence were less reliable than face-to-face encounters, especially when the accurate determination of personal identity or class membership was crucial to the transaction at hand. Despite the artifactual efficiencies of electrical media so admired by professional experts, they, too, attached greater weight to the irreducible face-to-face encounter as a more trustworthy guarantor of integrity. The less immediately present telephonic voice took longer to win social confidence.

New forms of presence muddied social distance. The most extreme fears were expressed as anecdotes in which protagonists who behaved as though full and attenuated presence could be treated as socially equivalent were promptly victimized by fast-talking, silver-tongued predators who knew how to take advantage of the fact that they were not. More or less constant anxiety was also directed to the effects of new forms of presence on customary social exchange, especially the practiced amenities of daily life.

> "When a man tells you a story face to face," said Horace C. Du Val, who is frequently annoyed by the telephone maniac, "he can see by the expression on your face, if he has the least knowledge of physiognomy, how the story strikes you, and it is an easy matter to cut a man off by a look or a gesture. But where are you when the story teller is 10 miles away? He has you cornered and you must listen."[60]

In the face of technological complexity, did the old proprieties apply, or did circumstances call for new ones to keep the social order intact? "To the woman who knows how to do things correctly," wrote *Telephony* in 1905, "it is positively maddening to have invited guests 'call her up' at a late date and acknowledge the receipt of her invitation and either accept or regret it. Especially nerve-trying is when the call comes in the middle of the dinner to which the person was invited."[61]

How properly intimate was a telephone? What line demarcated the social circumstances in which it constituted sufficient presence and in which it did not? Redefinitions of sufficient presence took a variety of forms. Some were simply procedural. At the annual meeting of the National Telephone Exchange Association in 1889 in St. Paul, Minnesota, a speaker related "how in Boston they managed to hold a directors' meeting with the Chair in New York and the Chairman in Boston, and how they would carry on their meeting and the secretary would record the Chairman as being present. The important thing was the voice; that being present they could hold a meeting."[62] Other redefinitions were legal and civic. Unable to appear before a police court in the matter of a breach of a bylaw, a Toronto citizen telephoned to court headquarters, admitted his guilt, and was fined a dollar and costs through the same medium.[63]

A more worrisome problem was how to defend social distances customarily enforced and maintained by face-to-face cues in the telephonic and telegraphic absence of these. Simply put, new media provided opportunities for the wrong people to be too familiar. Physicians were among those whose indignant complaints of telephone abuse received regular attention in electrical journals. According to a Chicago doctor:

> I give, every day, dollars and dollars worth of advice. If I sent a bill, the patient would probably faint, but she would be sure to call another doctor to bring her out of it and to attend her afterward. People who are ill send for us at once and then call us every day, often several times a day, and ask all the necessary questions for treatment. The telephone has become a really serious problem in the profession.[64]

Rules of presence were enforceable only when transgressors could be identified. The most disturbing assaults on social distance exploited telephonic anonymity. The *Leavenworth* (Kansas) *Times* reported that "disreputable" persons were nightly phoning "respectable" people and "using indecent, vile and vulgar language, and when asked where they are reply, 'at the *Times* office.'"[65] "There is a rule in all well conducted telephone exchanges," wrote *Electrical World* in the same vein in 1884,

> that when a subscriber, or any other man, is found swearing through the wire, promising his wife "Hail, Columbia!" if dinner is not on time

when he gets home, or otherwise abusing the facilities science affords
him for abuse of somebody else at a safe from kicking or hair pulling
distance, that subscriber is shut off at once. But there is little protection
against the voice of the swindler.[66]

Claims that personal moral qualities were lacking in telephonic
exchange, on closer examination, were complaints that missing and
missed cues were those of class-based deference. "Who cannot remember," wrote the *Electrical Review* in 1889, "when the telephone
was put into commercial use, being sometimes addressed by an unseen, and often unknown speaker, in language such as a man would
rarely use face to face with another man." One could determine, the
Review went on, who was a stranger to telephonic intercourse "by a
certain relaxation in the common courtesies of speech," as though the
unseen connection allowed callers to take refuge from the rules of proper
respect in personal invisibility, or remote presence. "There is an im-
personality," the *Review* continued,

> connected with the act of speaking by aid of a mechanism which has
> been lowering in its tendencies, for it is only by the personal presence
> of others that many men's speech is kept in restraint. On the other hand,
> a man of true refinement can usually be told by the courteous and grace-
> ful language which he uses when at the transmitter.[67]

A related problem was whether long-standing rules of propriety
that made it unthinkable for a gentleman to swear in the immediate
presence of ladies entitled him to speak more freely in personal telephone conversation with male friends. According to polite opinion, the
implied or actual presence of women, both the all-monitoring feminine
"Central" and the equally strong feminine presence at the center of the
family community increasingly served by the telephone, set the tone
for every conversation. An Ohio telephone company enforced its rules
against "improper or vulgar" language in phone communications by
removing the instruments of subscribers who did not observe this rule
of social presence. When a subscriber took the company to court on
the issue, the judge ruled: "The telephone reaches into many family
circles. . . . All communications should be in proper language. Moreover, in many cases the operators in the exchanges are refined ladies,
and, even beyond this, all operators should be protected from insult."[68]

Concern about profanity over the telephone was part of a larger
class of anxieties about how to speak properly in its presence, and what
community of speakers was addressed in the reach of its wires. Popular
periodicals encouraged readers to use the cautious good manners of

middle-class intimacy—quiet voices, clearly enunciated words, dignified presentation. "You speak into a telephone loud and harsh and you get a jarring sound, you misunderstand yourself," explained a speaker at the National Telephone Exchange Association convention in 1889; "but speak low and tender and you get a perfect communication of soul."[69] One writer felt that "something is gained in the cause of civilization that hundreds of thousands of persons all over the country are learning that tones sweet and low" were the most audible. "We have observed," he continued, ". . . that those who use the telephone have very much bettered their articulation and enunciation of words." Not only the diffusion of middle-class standards of speech was advertised as a happy consequence of the spread of the telephone, but also "the value of the habit of attention." Subscribers, while connected, must be "all ears and memory. . . . The mind cannot wander."[70] From a *Globe-Democrat* reporter who visited the operating room of the St. Louis telephone exchange, and was distressed at the way subscribers berated the "servants of the public," came the following admonition:

> There is one way to talk through a telephone, and only one. . . . That way is to stand back just a little and talk in an ordinary tone, as if you were speaking to a man a couple of feet away instead of to one three or four miles away, over a wire. Don't yell; don't whisper; simply speak in an ordinary tone and distinctly. The words are carried by electricity, not by the force of your enunciation. The instrument is not to be compared to a deaf man, but, to the opposite, is exceedingly sensitive.[71]

These instructions for speaking to public "servants" also suggest a concern for preserving class distinctions where few of the normal markers for signaling them were present. In some communities, competing telephone exchanges were differentiated by the social groups to which their subscribers belonged. "Like everyone else in Tuxedo, when the telephone first came into vogue we used it continually as a method of conversation with our friends," reported a society lady from New York in 1899, her notion of "everyone" an index of the limits of her world.[72]

If some cues were unfortunately missing, others were in danger of being revealed. The same society lady related how the phone in her husband's dressing room rang one night after she was in bed. Rushing to answer it, she fell in the dark into a half-full bathtub. The caller was a friend telephoning from "the club" with a large party of visiting friends.

"'Charlie S—— wants to speak to you.' Then I heard laughter and chaffing, and I could fancy them all in the gay clubroom in their smart dresses. In a minute, Mr. S—— spoke. 'How do you do, Mrs. A——,' he began, and for fully 10 minutes I had to carry on a light conversation, first with one and then with another of the party, while I stood shivering in the dark in my damp and scanty garments."[73]

The shame of not being publicly presentable, as though one were not actually invisible and safe at home, was the subject of uncomfortable humor. At a large dinner at a "Fifth Avenue palace" in 1899, reported the *New York Tribune,* an absent-minded guest was missing. Her hostess telephoned impatiently.

"We are waiting dinner for you, Amelia," said the magnificently gowned woman at one end.

"Good gracious, Margaret, is this the date I was to have dined with you?" exclaimed Amelia, who was in a dressing gown and slippers, at the other end. "I am simply overwhelmed with mortification, but I entirely forgot that this was the night."

"Jump into a cab and come as you are," urged the first speaker.

"You wouldn't ask me to if you could see me!"[74]

To those persons, usually men, whose rebellion against the social order took the form of distaste for rigid dress codes on public occasions, exemption from visual accountability in the invisibility of telephonic connection was welcome. Anticipating the "telephone banquet" of the future, a comic writer in *Telephony* asked, "Why hasn't some one remarked that the new plan proposes to annihilate that abomination, the reception committee, of Prince Albert coats? That is the sort of thing we mean by 'progress.'" With a telephone in his room, an orator might "sit in dressing gown and slippers and talk at ease to comfortable minded listeners. . . . Not only does he evade ruining his digestion with countless 'banquets'—he also saves his temper by not having to shave."[75]

The much larger number of stories that expressed fear of being seen over the telephone does give an indication of the way in which the psychological novelty of telephone conversation was layered uneasily over social procedures of a traditionally established kind. The discomfort that accompanied so apparently trivial a readjustment of social thinking suggests how many more times that response was magnified in response to intrusions much closer to the heart of community identity.

Crimes of Confidence

The association between sensational crime and new electric media was strong in popular and expert literature. Several sturdy crime anecdotes were part of an enduring popular tradition about the early social history of both telegraphy and wireless. Two stories in particular were credited with presenting these media to popular attention around the time of their introduction. In one, a description of John Tawell, who murdered his mistress in Slough in 1844, was wired ahead of him as, disguised as a Quaker, he fled by rail to London, where the authorities soon spotted him. The device of his undoing was immediately and popularly labeled "the cords that hung John Tawell."[76] In the other story, also British, Dr. Hawley Crippen murdered his wife in 1910 and buried her body in a coal cellar. He booked passage on the S. S. Montrose with his mistress, masquerading as his son, and was apprehended as a result of messages sent by wireless to Scotland Yard from the captain of the ship, who suspected Crippen of being the wanted man.[77]

"It is a well-known fact that no other section of the population avail themselves more readily and speedily of the latest triumphs of science than the criminal class," explained Inspector Bonfield to a *Chicago Herald* reporter in 1888. "The educated criminal skims the cream from every new invention, if he can make use of it."[78] The article went on to recount a telephone crime story, said to be entirely factual, that represented a substantial genre of stories about the use of the telephone by con men to deceive those accustomed to conducting their business and social affairs in person:

A millionaire speculator at the Chicago Board of Trade had a phone line installed between his office and his home, one of the first such lines in the city. One day he answered a call from a man who claimed that he had bound and gagged the speculator's wife and servants and set oily rags around the house. The businessman was instructed to give twenty thousand dollars from an office safe to a confederate waiting at that moment outside his office—or else. The ransom delivered, he rushed home to discover his wife serene and tranquil, the family hearth undisturbed. His gracious helpmeet had invited into the house a well-dressed gentleman who asked to make a private business call to her husband. "He used the telephone but a couple of minutes and then came back into the parlor, thanked me for the favor I had done him and went away, saying the telephone was a very useful invention." The use of new media to serve personal ambition, to achieve status, success, and social control, did not guarantee that the same powerful tools could not be turned against one, and one's family and oneself made desperately vulnerable in the bargain.

If the lone woman at home seemed especially vulnerable to predators, she could also lift the telephone to sound the alarm, in many stories a device by which help was quickly dispatched to thwart thieves, murderers, and frustrated suitors.[79] A housekeeper surprised by burglars in Bristol called "Central," who advised the police, who soon had the burglar in custody. The incident occurred at the exchange of the Western Counties and South Wales Telephone Company, which published a circular advising "timid householders of either sex, fearing fire or thieves" to connect at once to the local exchange.[80]

Women were considered especially susceptible to male manipulators of electrical technology because of their less-worldly experience in gauging trustworthiness. Their experiences were assumed to be limited to intimate, orally based communities where close and constant association discouraged the kind of deception that was possible in electrically constructed communities where unknown parties could pretend a dangerous familiarity. Widows and lone women were particularly helpless. A Milwaukee widow met a confidence man representing himself as a grain broker from New York. Courting her by letter from St. Louis, he soon wrote that he had been summoned to Europe on business. As they would be unable to meet again before his departure, he proposed that they should marry in a telegraphic ceremony.

> An ecclesiastical crank of Chicago lent his aid to the blasphemous travesty on wedlock, and the lovers were "married" by wire, the bride sending him, she said, "an electric kiss." His reply came from England in the shape of a request for a speedy remittance of $2,000 to help him in a business transaction. This was sent him and was followed by another request for $1,000, which sum was also sent, after which all trace of him was lost.[81]

Because her suitor followed impeccable social form in one mode and courted her by letter in the acceptable long-distance way, the Milwaukee widow assumed his behavior in another mode would be a faithful analogue. Perhaps, too, she was seduced by the romance of high technology—not the first woman so deceived. The penalty: loss of love, loss of property, loss of face and reputation—in short, loss of the principal credentials for membership in the social order to which she belonged.

A more sensational example was the court-ordered annulment of the marriage of Sarah Orten Welch and Thomas Welch in Indianapolis in 1883. While a resident of Pittsburgh, the plaintiff, Mrs. Welch, answered a "personal" advertisement in the *Cincinnati Enquirer* and struck up a correspondence with the defendant. The two correspondents were duly married, not in the customary face-to-face way, but by telegraphic ceremony. When the new bride, from whose point of view the story was told, went to meet her husband, she found that he was "a colored man" and a barber.[82] Since Sarah Welch had corresponded with her intended husband by mail as well as by wire, the printed medium had betrayed her no less than the electric, but public attention focused on the "telegraphic farce of wedlock." This had too lightly replaced the oral sacrament whose minimum condition was the physical presence of the principals at the marriage ceremony, where each could observe the other and be observed by the community, whose order was at stake in every connubial alliance.

An exception to the general social proscription against telegraphic wedlock, predictably, was romance within the established fraternity of telegraphic professionals. In 1876 the *San Diego* (California) *Union* featured the story of a telegraphic wedding between Clara E. Choate of San Diego and W. H. Storey, U.S. Signal Service operator at Camp Grant, Arizona.[83] At the last minute, operator Storey was denied leave to go to San Diego for the appointed ceremony. There was, further, no qualified person within hundreds of miles to perform the nuptials at Camp Grant. Miss Choate went to Camp Grant anyway, and the services of the Reverend Jonathan L. Mann, pastor of the Methodist Episcopal Church of San Diego, were telegraphically secured. Lieutenant Phillip Reade, the officer in charge of government lines for California and Arizona, cleared the line for 650 miles from Camp Grant to San Diego for the unusual ceremony, and issued a general invitation up and down the division inviting all station managers and their guests to be telegraphically present.

> Each operator on the line accepted this novel invitation and with invited friends was present at his station as wedding guests. At 8:30 p.m. the father of the bride sent this message from San Diego to the wedding party at Camp Grant:

> "Greetings to our friends at Camp Grant. We are ready to proceed with the ceremony.   D. CHOATE AND PARTY"

> The answer at once came back:

> "We are ready.   W. H. STOREY
>   CLARA E. CHOATE"

As Mr. Mann read the ceremony, Chief Operator Blythe at San Diego transmitted it word for word to Camp Grant. The ceremony concluded, the chief operator wired the newlyweds that "the Silver Cornet band of San Diego is just outside the office, giving you and your bride a serenade." In later years, it was not uncommon for the bridegroom to meet perfect strangers who would exclaim upon learning his name, "Storey! Let me see, wasn't you married by telegraph some years ago? . . . Well, well, I was a guest at your wedding. I heard the whole ceremony at the telegraph office at ————, Arizona." This telegraphic wedding had what others lacked: participation by community members in good standing, both the regulars at any properly conducted wedding, and the brotherhood of telegraph professionals, all acquainted. Instead of narrowing the community of moral accountability, the telegraph had for a brief moment enlarged it.

Dramas of wrongdoing that exploited ambiguous presence often turned on the subtle drama of proscribed relations between persons of unequal social status. Cases of telephone fraud were often most successful when their perpetrators pretended to rank high in the chain of social command. This was because any social inferior who challenged a legitimate claim of respect due a social superior risked heavy penalties for disregarding the rules of social distance. A clerk at the Ebbit House, a leading society hotel in Washington, D.C., was deceived when a confidence man took advantage of the social insulation between the upper-crust clientele of the hotel and those in their service. The clerk received a call from someone who claimed to be "Major Peabody, a well-known permanent guest of the house," who instructed the clerk to pay charges of $54.50 on a suit of clothes that was soon delivered as announced. The fraud was discovered when the real Major Peabody arrived for dinner.[84] In Detroit, an elegantly attired gentleman fleeced a bartender for a gin fizz and an imported cigar by pretending to phone up a prominent local businessman to cancel an important appointment, "discovering" suddenly that he did not have enough cash for his refreshments, and promising to come back the following day to pay up. "The barkeeper didn't kick at all, the telephone conversation had fixed it," and the successful con artist never returned.[85] Because Erastus Wiman, owner of the ferry boat line between New York and Staten Island, occasionally called the pier with instructions to hold the last boat of an evening for his arrival, several enterprising young men tried phoning the pier and pretending to be Wiman in order to delay the boat until they could get to it themselves.[86]

Sometimes those lower on the social scale did mischief by ap-

propriating a communications technology their social superiors regarded as exclusively their own, and which they could not imagine
those lower in the social order being clever or impudent enough to
resort to; or perhaps they were unable to imagine people socially unlike
themselves using the telephone, since they assumed that such people
would have no friends with telephones, and no one to talk to. William
J. Brown, "a disgruntled laborer" discharged from the Brooklyn lum-
beryard of Alexander & Ellis, did the unimaginable. He telephoned
his former employers and ordered a large quantity of lumber in the
name of a regular customer. Next, he intercepted the delivery and had
the driver dump it in a vacant lot, in exchange for a forged receipt.
This neatly sewed up the deception of his employers, who also trusted
literate procedures. Finally, he carted off the lumber to sell it at a
profit.[87]

In the conflict between high and low, high was not always right,
however—at least not altogether right. *Comfort* magazine printed a
fiction story about John Drummond, an electrical stalwart against silk-
stocking crime, "supplementing his 'Tech' education with a year of
practical work . . . in the employ of one of the great electric companies." In the course of repairing phone lines after a storm, Drummond connected his movable telephone to a random dangling wire.
Fate, manifested in unusual weather that caused mysterious electrical
phenomena, rewarded John Drummond for performing his professional
duties in a conscientious manner. He heard, conversing near an open
telephone at the other end of the wire, an ungrateful son plotting to
steal five thousand dollars from his wealthy father. This information
was delivered to the father in a timely fashion, and the crime was
aborted. The story celebrated the morality of the young professional
in training who had so discreetly preserved the social order. The rich
man kept his wealth and, as the story noted, nothing appeared in the
papers. At no compromise of his principles, the electrician had proved
indispensable to the maintenance of both public order and class structure.[88]

In urban environments, one could not be too careful with strangers
who might falsely represent themselves as insiders. Familiar neighborhood circles could be infiltrated even by legitimate insiders with a
false face. Typical was the 1907 tale of a suburban couple staying
overnight with friends. While telephoning a neighbor with instructions
to lock up their house, they heard the click of another party lifting a
receiver on the ten-party line. This "made no particular impression, as
it was so common an occurrence." When they returned home, how-

ever, they found their house ransacked and robbed. According to the Baird Manufacturing Company, which claimed to have a system to protect householders against such invasions of home and privacy, police investigation had "developed the fact that a great many robberies had been planned and executed in the city from information obtained by 'listening in' on party lines."[89]

Further electrical duplicities were reported. A racing jockey in Guttenberg, New Jersey, concealed an electric battery in his silk jacket, connected to wires running down his legs to the spurs of his boots. By this means, he shocked his horse on to greater effort. He was discovered only because he weighed nine pounds extra as a result.[90] The Lewiston, Maine, police found their liquor raids frustrated because a system of electric alarms operated by the saloons allowed outside "loungers" to press an electric button that warned the bartender inside of the arrival of the law.[91] Chinese gambling dens in San Francisco protected themselves from police raids by concealed electric floor mats that set off special alarms at the approach of intruders, and by backroom electric alarms triggered by sentinels posted at front doors.[92]

Defending Nondemocratic Communication

The conviction that new technologies of communication could help fight crime was no less popular than the idea that they would encourage it. There was no aspect of criminal surveillance or police intelligence to which their application was not suggested. Electric devices to foil or frighten criminals, gather evidence, or alert authorities to crimes in progress and citizens in need were the subject of elaborate and excited speculation. So were proposals to use secret photography to catch burglars in the act.[93] Another invention that could be "pressed into the detective service and used as an unimpeachable witness" was the phonograph. "It will have but one story to tell," explained *Scientific American*, "and cross examination cannot confuse it."[94] Machines that moved messages rapidly across space could relay information about criminals and their movements to centrally located authorities able to direct pursuit. In stories with a future dimension, wireless telegraphy was often enlisted. "The timid householder," mused the popular British journal *Answers* in 1899, "will henceforth keep a bottle of wireless telegraphy by his bedside, and flash a silent appeal for help to the nearest police-station when treated to a surprise visit by the ubiquitous and undesirable burglar."[95]

Police departments were one of the earliest groups of telephone users. Boston police stations had been fitted with telephones by 1878.[96] By 1888 Baltimore had established a system of 170 call boxes throughout the city to keep policemen on the beat in touch with headquarters. Each sergeant was required to telephone his station once an hour.[97] A variety of electrical devices protected valuable property. Electric burglar alarms were considered a marvel of the age. Some included photographic devices to record suspicious activity. Others shocked intruders on the spot, or flooded the premises with light to scare them away. Special canes and umbrellas concealed electric devices to fend off robbers on the street. The finest treasures of the Metropolitan Museum of Art were protected by an electric warning system connected to the director's office.[98]

Electric protection of property was considered an indispensable aid to public order, but the moral force of the electrician's calling was promoted more enthusiastically still in the idea that electricity provided the connecting links for a political order that sustained the economic one. "Suspend telegraphic communication for a day, and the uncertainty would spread distrust in financial circles, silence the fire alarm for a night and any one of our combustible cities might be known by its ashes alone, and extinguish the electric light while the sun is beneath the nadir, and crime would riot," a prominent electrician declaimed to the New York Electric Club with characteristic rhetorical extravagance.[99] Electricians and politicians instinctively reached out to one another, even though relations between them were not always smooth. Electricians recognized the state as a prospective ally and patron, and the state recognized the utility of electrical technology, especially electrical communication, for expansion and control. This was the thrust of a statement by the British Special Commissioner Henry Norman in 1888 in honor of a demonstration of the longest telegraphic circuit ever worked, an 8,100-mile connection from New York to London through Canada, the occasion for which was the commissioner's visit to Vancouver: "Is not the click of this key—heard in two hemispheres—more eloquent than all the arguments of empire ever penned?"[100]

National governments believed sufficiently in the value of electrical control to spend increasing amounts on it. The *Electrical Review* of London reported that the Chancellor of the Exchequer had put the total cost of telegrams from all government departments during 1884 at roughly £16,000 sterling. Subsidies amounting to £49,000 net had been paid by the Imperial Exchequer to submarine telegraph companies in the previous year. When the Eastern Telegraph Company laid two cables to Australia, one was subsidized by the Australian colonies at £32,400 annually.[101] In 1896, William Edward Hartpole Lecky wrote of the telegraph:

> It has brought the distant dependencies of the Empire into far closer connection with the mother country; but it is very doubtful whether the power it has given to the Home Ministers of continually meddling with the details of their administration is a good thing. . . . The telegram, on the other hand, has greatly strengthened the central Government in repressing insurrections, protecting property, and punishing crime. It has, at least, modified the Irish difficulty by bringing Dublin within a few minutes' communication of London.[102]

Prince Bismarck, wont to complain of the burden "of being kept incessantly on the alert by electrical communication," was connected directly by wire with the foreign office at Vienna, and "with its assistance he controls Austro-Hungarian politics."[103] "One of these days," predicted *Electrical World*, "the office of the Secretary of State at Washington will be connected by wire with the foreign offices of Mexico and Nicaragua, for the better discharge of our duties of 'protectorate' over those countries."[104]

In their appointed task of approximating ideal cities, industrial expositions deployed electric communication as a form of efficient social control. Typical was the fire alarm and police telegraph system set up on the grounds of the 1885 World's Exposition in New Orleans. "The buildings are constantly patrolled by thoroughly disciplined policemen, who on the slightest approximation to fire, undue heat or disturbance of any character anywhere, will immediately send the alarm."[105] During railroad strikes in 1892 at Buffalo, New York, the telegraph and telephone were "invaluable aids" in summoning military personnel from their homes, clubs, and vacations.[106] "In view of the riots and unruly meetings of the unemployed which now take place every winter in the West End of London," reported the *Electrician* in 1888, the London police had connected all stations by a complete and independent telephone system. Additional direct lines connected high police officials and the fire brigade, the houses of Parliament, and the War Office. "It will be possible to collect a large force at any desired point, within a very short time."[107]

Electric control could also be exercised over underlings in the workplace. A British observer reporting on "daily practical" applications of electricity in America wrote that some American factories used electric "tell-tales" to monitor night watchmen.

Keys are placed all over the building, to be depressed by the watchman when he goes his rounds; this operates a registering pencil in a clock. The register-sheet, which is changed daily, and which is placed in the manager's office, cannot be got at by the watchman. The man's whereabouts at any hour or minute is thus permanently registered, and can be checked in the morning.[108]

The most powerful demonstration of social control through electricity was no exercise in communication, however, but in ultimate, unanswerable force. The death penalty, *Scientific American* speculated in 1876, would be a more effective deterrent if electricity were used to administer it, as "the peculiar death by lightning, which, among the ignorant of all nations and ages has been the subject of profound superstition, would, without doubt, through its very incomprehensibility and mystery imbue the uneducated masses with a deeper horror."[109] *Answers* predicted a no less total but more subtle control of citizens by 1950 through a medium of mind control whose characteristics of action at a distance were suggested by electricity and embroidered in a typically futuristic way:

If any person is drunken, lazy, or diseased, he is immediately cited by hypnotic summons—the telephone is now obsolete—to the municipal laboratories to receive the proper microbes to make him a decent citizen. If he fails to attend, the operator (a powerful hypnotizer) brings his influence to bear upon him, and he then comes without any further difficulty. So policemen, gaols, workhouses, etc., are all abolished.[110]

The choice of electricity to defend a despotic order of force over a democratic order of mutual communication suggests a system of political values dramatically at odds with public rhetoric about the socially harmonizing effects of electric communication, and with promises, albeit lofty and vague, to make it more democratic, and soon. This rhetoric created a dilemma for those in charge, or aspiring to be in charge, since a truly democratic communications network would cease to be an instrument for excluding or segregating citizens by rank, in a class community that cared very much for such distinctions. At different moments, electric communication was presented as a herald of democracy *and* as a superior means for controlling masses, criminals, primitives, servants, and whatever other underclasses might need restraint. The literature of egalitarian exhortation in behalf of a future of universal electrical consumption also printed subtle and not so subtle arguments as to why the telephone, the most sophisticated instrument of electrical communication, would never become truly democratic.

One line of argument presented the social telephone as a luxury play-
thing noxious to a domestic life of sober responsibility. This was the
suggestion behind a statement attributed to a British aristocrat in *Lightning* in 1893:

> [The telephone] was tried recently in a large country house as a brilliant
> idea to relieve the sombre function of entertaining Royalty, and suc-
> ceeded passing well. But possibly, when the novelty has worn off, the
> amusement will vanish also. Personally, there are few things I dread
> more than using a telephone.[111]

When the poor were portrayed as speaking about their own interest in new electric media, the sentiments attributed to them were
often slyly framed by the suggestion that this was a world far beyond
their conception. This joke from *Electrical World*, replete with the
debased language that characterized the poor from the perspective of
literate privilege, made that point:

> "'Lectric wires, indade! Arrah, they'll make foine clothes lines for
> us poor folks when they rin 'em over these roofs; they're just clane and
> purty for it."
> "Och! bad cess to yes if yer tech wan o' thim. Me ould man sez
> as how one ind on 'em's fastened to the divil's own tail and the ither
> rests in the furnace o' hell. Yer'd be half way through purgatory afore
> your washin' was dry."[112]

Arguments were made that the utility of the telephone could not
be preserved without restricting its availability. Keeping British telephone service expensive and exclusive expressed political priorities about
whose communications needs were important to the British government and whose were not, as in this statement by Postmaster-General
Arnold Morley to Parliament in 1895:

> the telephone could not, and never would be an advantage which could
> be enjoyed by the large mass of the people. He would go further and
> say if in a town like London or Glasgow the telephone service was so
> inexpensive, that it could be placed in the houses of the people, it would
> be absolutely impossible. What was wanting in the telephone service
> was prompt communication, and if they had a large number of people
> using instruments they could not get prompt communication, and yet
> make the telephone service effective.[113]

Speaking to the Telephone Convention in Philadelphia in 1888,
Dr. S. M. Plush argued that misguided zeal in regulating telephone
rates had induced those with "no real need for the facilities" to become

subscribers. Nothing, he felt, had disadvantaged telephone companies and their subscribers more than "cheap rates." As a result:

> This lends a tone to their business, it is true, and serves in a measure as a card of respectability. Their place at once assumes an imaginary importance, and if there is any one thing that delights them more than another, it is having some real or fancied business with a leading and substantial firm. This they manage to have at frequent intervals, thus taking the time and monopolizing the wires of others to their detriment and absolute loss.[114]

An exchange could be so large as to be worthless. In the interests of "busy houses," a time might come when the price of an annual rental fee would be an insufficient method for weeding out undesirable subscribers, making it necessary to implement a method whereby suitable subscribers would be required to hold membership certificates that entitled them to service.

The idea that the telephone was a general-purpose communications device instead of the exclusive property of a small group that used it for narrowly defined ends had to be invented and defended, and had to do battle in the course of its development with a far more limited notion of telephone possibilities. It happened that certain indignant New England subscribers gave up their phones after rates rose from $30 to $36 a year for residences and from $36 to $42 for businesses. In response, the editors of the *New York World* sniffed, "We maintain that if a subscriber has no more use for a telephone than is represented by an outlay of $42 per annum, he evidently does not need the telephone at all."[115]

Confrontations between the restricted elite telephone and the inclusive convenience telephone were sometimes dramatic. In Britain in 1889, postal officials reprimanded a Leicester subscriber for using his phone to notify the fire brigade of a nearby conflagration. The fire was not on his premises, and his contract directed him to confine his telephone "to his own business and private affairs."[116] The Leicester Town Council, Chamber of Commerce, and Trade Protection Society all appealed to the postmaster-general, who ruled that the use of the telephone to convey intelligence of fires and riots would be permitted thenceforth.

The conflict between these two visions of the telephone was sharpest when the issue at stake could be identified as a critical public interest. A committee of the National Telephone Exchange Association discovered that local fire departments differed among themselves about whether telephone alarms called in by the general public should replace or supplement official telegraph alarms. The committee conceded the existence of more telephones than telegraph alarm boxes, the greater speed of telephonic communication, and the fact that telephones were already used to communicate between fire department headquarters and enginehouses. It had documented "hundreds" of cases in which telephone alarms had been effective. Still, "in almost every branch of business" there was concern that popular channels of communication could not be relied on for reporting emergencies. One fire superintendent feared that if subscribers could easily reach his office, "he would be constantly annoyed by false alarms and trivial questions and that his department would be burdened with a great amount of unnecessary work and damage."[117] On the other hand, a case was reported to the committee in which a woman alone in her house

> called up headquarters and stated, very coolly and explicitly, that the house was on fire and that there was urgent need of the assistance of the fire department which she could not, without difficulty, reach in any other way. As this was not according to the "*official*" routine, the operator declined to give any alarm and remained quietly in his office for about ten minutes until the alarm was received from a . . . box, quite distant from the burning building, which, in consequence, was almost entirely destroyed at a loss of not less than fifty thousand dollars.[118]

Proposals to widen the reach of telephone service were often indignantly resisted. In 1884 a subscriber in Edinburgh was outraged to learn that his local phone company planned to put a number of telephones "where any person off the street may for a trifling payment— a penny is suggested in some places—ring up any subscriber, and insist on holding a conversation with him." Against the plan, he argued that

> subscribers have the security at present that none but subscribers can address them in this way, and that these are equally interested in the telephone not becoming a nuisance. But if everybody who has a penny or threepence to spare can insist on being listened to by any of the leading business establishments of the city, we shall only be able to protect ourselves against triflers and intruders by paying less regard to all telephone communications.

By an unusual logic, he concluded that more accessible telephone service would cause the number of subscribers to drop. The presence of

fewer subscribers would return the system to equilibrium and its previous state of inaccessibility, but with a different set of subscribers,
doubtless of a less desirable complexion. Subscribers of social substance would find their special preserve overrun. People like him would
be forced to publish notices next to their telephone numbers advising
that they received calls only at certain hours. Perhaps they would have
to hire "telephonists" to sift verbal communications, just as secretaries
weeded correspondence.[119] Such efforts to recover the exclusivity of a
more restricted telephone service were doomed to fail. The only defense against the widening democracy of the telephone was to keep
one's name out of the directory, and by 1897 it was claimed that many
busy men did this.[120]

A similar plan was put into effect in Glasgow in 1888 with sev-
enty-six automatic call boxes. Subscribers were issued keys, while
nonsubscribers were required to deposit three pennies or six, according
to the distance called, for a three-minute connection, which was broken after the elapsed time.[121] In Paris callers were issued fifty-centime
(ten-cent) tickets to go to a post office for a five-minute conversation
with any other telephone in the city. In addition, telephonic communication was established between drugstores and hospitals by the Society of Ambulances, for emergency use in case of accidents.[122] In
1889, the New York and New Jersey Telephone Company made plans
to erect "slot-telephone machines" at elevated stations and other places
of "public resort," where for a nickel patrons could ring up Central to
complete the connection.[123]

The accessibility of the telephone was an issue in a number of
American court cases that tested whether subscribers whose friends or
customers used their telephones were liable for more than the usual
flat rate, or indeed could have their phones removed. As early as 1885
a subscriber sued the Southern Bell Telephone Company for five thousand dollars after the company cut his line as a penalty for allowing
nonsubscribers, or "deadheads," to use his phone. A jury in Charles-
ton, South Carolina, found for the plaintiff, awarding one dollar and
costs. The judge had instructed that nonsubscribers could not avail
themselves of the plaintiff's telephone by the terms of his contract, but
added that the company could not cut the wire without having forbidden the practice in writing.[124]

In 1898 the electrical press picked up a story in which Mr. Danenhower, proprietor of the Fredonia Hotel in Washington, D.C., secured an injunction against the Chesapeake and Potomac Telephone

Company for attempting to remove the telephone that had graced his hotel lobby for eight years. Danenhower had repeatedly ignored company warnings against allowing hotel guests to use the instrument. "On January 14th, after I had used the telephone four times, a guest of the hotel used it, and when he had finished, Mr. Bryan called me up personally and told me if I didn't cease the illegitimate use of the telephone it would have to be removed."[125] He was not even permitted to call his warehouse, which had boasted its own instrument for thirteen years, to inquire about ill family members without paying a ten-cent charge for each call over and above the regular annual rental fee of $100 for his hotel phone, $120 for his warehouse phone.

In U.S. District Court, Justice Cox stipulated that Danenhower must observe the terms of his contract with the telephone company. Guests could not use the instrument for personal business. Use of the telephone must be limited to the "benefit and the accommodation of boarders," which included sending for a wagon or carriage, but precluded calling up for stock reports or ordering theatre tickets. The judge ruled that the telephone company must serve all patrons on the same terms, and he suggested that the telegraph provided the relevant model. Though telegraph companies often placed instruments in hotels, hotel guests did not expect to use them personally. "It is very annoying," he wrote,

> for a subscriber to call for a certain number and be told that the connection he wants is busy, that being due to the fact that a non-subscriber is using, free of charge, the telephone with which connection is desired. . . . The guest does not have his telegrams sent free of charge, nor does he secure the service of messengers and the use of carriages without pay. Why should he be allowed to use the telephone without paying for its use?

One of the telephone company's objectives in the Danenhower case had been to create a demand for dime-in-the-slot telephones in hotels, drugstores, and police stations, where available phones were often treated as community message centers by nonsubscribers.[126] This solution was arrived at after druggists, hotel proprietors, and others repeatedly resisted telephone company efforts to raise rates for commercial subscribers whose telephones were available to the public or their own clientele. In 1891, for example, three hundred druggists in Baltimore signed a petition threatening to order their telephones removed from their stores in the event of a proposed increase in rates for messages transmitted by anyone other than themselves as original renters.[127]

The nature of the public demand for telephone services was varied. In 1888, a Chicago druggist explained to a newspaper reporter the range of messages he was expected to deliver daily. When the circus had come to town, for example, a gentleman had requested him to inform the gentleman's mother, who lived two blocks away, at what hour the parade would pass his office. A young newlywed persuaded him to call up her husband at a downtown store. She took the receiver and launched into a lengthy and intimate lovers' conversation. Another young woman asked him to deliver to a certain young man in the neighborhood the message that she had decided not to marry him and all the wedding presents would have to be returned. Another man had awakened him in the middle of the night begging him to telephone the police to report a lost child. Still another man had prevailed on him to call a physician who lived three or four blocks away. He resented this monopoly of his time without compensation, but conceded that drugstores had been providing such services for years, and the public now expected it.[128] He paid $150 a year for his telephone, and though some customers occasionally paid for its use, most did not.[129]

Employees at establishments with telephones were usually regarded as legitimate business users of those telephones, but the expansive temptations of the sociable telephone made narrow rules of conversation difficult to enforce, especially when the employees of the telephone company were wont to engage in a bit of sociable exchange. The *Brooklyn Times* described a kind of rule-bending that even the telephone company could not control:

> The joy of the all-night telephone girl, the one upon which her chief reliance is placed to help her cheat the long hours of weariness, is the sociable night clerks in the always-open drug stores. The latter entertain a sort of brotherly affection for the night operator, and evidently lose much sleep in inventing ways for her entertainment. The ways of the young drug clerk, fresh from college, are usually refreshing. As a general thing he can warble sweetly, and any knots that may be in his vocalization are toned down and smoothed out in transmission. His repertory is inexhaustive and includes everything in the song dealers' catalogue, and much that is not, from "Just One Girl" to "Cavalleria Rusticana." Then he tells such funny stories and is so unassuming. Besides, it is a poor drug store nowadays that doesn't boast a music box, and this is frequently brought into requisition for the benefit of the all-night operators.[130]

Despite energetic efforts to limit nonsubscriber use by threatening to remove instruments, by installing pay telephones, and by issuing tickets to nonsubscribers, to be surrendered to rightful subscribers call by call, emerging networks of telephone sociability undermined every effort to make the telephone restrictive and inegalitarian. "It is surprising that the number of people who use a telephone and never pay is so large," the *Electrical Review* mused. "The telephone is apparently looked upon as a public convenience, and quite often in the smaller cities a single telephone is expected to answer for an entire block."[131]

In 1900 the *Western Electrician* predicted that telephones would grace "every well-ordered household in the United States within the next decade." This would depend on substantial rate reductions, which its editors believed competition would provide. In the old days of the Bell patent monopoly (1879–97), the Bell Company had had "no need to popularize the telephone; it preferred a limited service at high prices to a widespread use of the telephone at rates within the reach of persons of moderate means."[132]

Thomas Edison had given his opinion of an economically democratized telephone service to the *Boston Herald* in 1885:

> "The tariff rates are now so high as to preclude the average householder from taking advantage of the service unless a number of subscribers are all connected with one wire. Of course, this is objectionable, as each subscriber on the wire hears whatever conversation is going on over it. Recent experiments have shown that all this may be obviated, and that one of a number of subscribers on the same wire may have his own words secure against eavesdropping. The perfecting of these experiments will lead to great extension of private house business, and the telephone will come within the means of the ordinary householder."[133]

With the advent of the telephone and other new media came relatively sudden and largely unanticipated possibilities of mixing heterogeneous social worlds—a useful opportunity for some, a dreadful intrusion for others. New media took social risks by permitting outsiders to cross boundaries of race, gender, and class without penalty. They provided new ways to silence underclasses and to challenge authority by altering customary orders of secrecy and publicity, and customary proprieties of address and interaction. Well-insulated communities of pre-telephone days could not remain forever untouched by these developments, nor were telephone companies able to ensure that emerging telephone communities would keep within the bounds of social decorum and work-related use. Somewhere between the expansive intentions of entrepreneurs and the practiced exclusivity of familiar social codes, the telephone and other new media introduced a permeable boundary at the vital center of class and family, where innovative experiments could take place in all social relations, from crime to courtship.

Artist's sketch of central display in the Electricity Building, World's Columbian Exposition. Mounted by General Electric Company and Phoenix Glass Works, Chicago, 1893. (*Electrical World*)

Mrs. E. E. Gaylord represents electrical enterprise at "The Greatest Event in the history of Brookings, South Dakota," Brookings opera house, 1890. (*Western Electrician*)

Mrs. Cornelius Vanderbilt as "The Electric Light," 1883. (*Courtesy of The New-York Historical Society, New York City*)

Westinghouse electric light exhibit at the Columbian Exposition, Chicago, 1893.
(*Electrical World*)

Edison float representing Electricity for pageant celebrating the 400th anniversary of Columbus's discovery of America, New York City, 1892. (*Special Collections, Van Pelt Library, University of Pennsylvania*)

Brooklyn Bridge welcomes Admiral Dewey, New York City, 1899. (*Western Electrician*)

Electric light portrait of Admiral Dewey's flagship, Olympia, Mandel Brothers' storefront at State and Madison Streets, Chicago, 1899. (*Western Electrician)*

French electrical jewelry, 1879. In circuit, the rabbit drums a gong, the bird moves its wings, and the skull rolls its eyes and gnashes its teeth. (*Scientific American*)

George W. Patterson and his electrical novelty act, Chicago, 1899. (*Western Electrician*)

George W. Patterson producing electrical spectacular effects with lighted clubs.

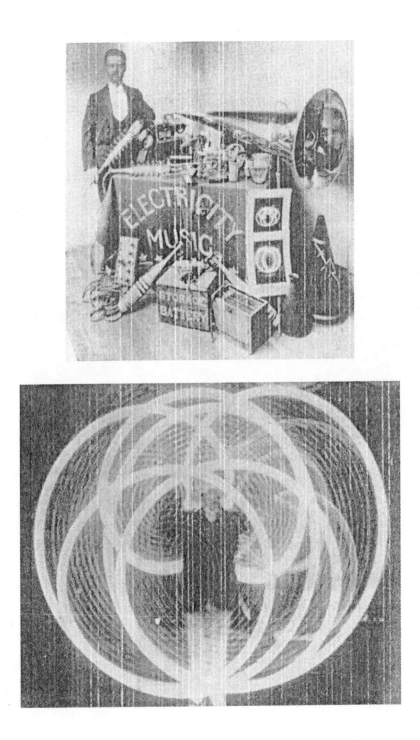

THE PERSONAL NUMBER
THE LADIES' HOME JOURNAL

FEBRUARY 1912
15 CENTS

Ladies' Home Journal cover from 1912 associating telephone use with class etiquette.

Experimental uses of new technology grace the cover of *Invention*, London, 1896. (*The British Library*)

A stentor reads the news over the Telefon Hírmondó. (*World's Work*)

Transmitting a musical program over the Telefon Hírmondó, Budapest, 1901.
(*World's Work*)

An evening at home with the Telephone Herald, Newark, New Jersey, 1912.
(*Literary Digest*)

3

Locating the Body in Electrical Space and Time
Competing Authorities

The young ladies of Frankford . . . have recently discovered that by holding a piece of tin against the iron foot-rests driven into the wooden poles of the Suburban Electric Light Company they receive a weak electric shock, and almost every evening a group gathers around the poles that are not situated on the main thoroughfares and enjoys the fun for hours . . . One pretty miss was heard to remark, after her first experience, "Oh, I thought I was squeezing a handful of pins." "Yes," said another, "it's something like being kissed by a young man with a bristly moustache."

—*Philadelphia Record*, 1891

In any culture codes for bodily communication are conventionally elaborated and, like other codes, require skillful manipulation. The body is the most familiar of all communicative modes, as well as the sensible center of human experience, which lives or dies with it. Upon it, all other codes are inscribed to a greater or lesser extent. There is no form of communication that does not require the body's engagement, though printed and written messages may involve a smaller direct range of its perceptual and motor capacities than oral-gestural messages do. In addition, strange experiences are often translated and made familiar by comparisons with the body, and by categories of classification derived from the body's experience. The body is a convenient touchstone by which to gauge, explore, and interpret the unfamiliar, an essential information-gathering probe we never quite give up, no matter how sophisticated the supplemental modes available to us.

The body is also squarely at the critical juncture between nature and culture. It *is* nature, or in any case man's most direct link to nature, capable of opposing and resisting it at least for a while, either its own or that external to it. The inscription of cultural codes upon the body is perhaps the principal means of detaching it from nature and transforming it into culture. The body and its actions, therefore, have a richly ambiguous social meaning. They can be made to emphasize perceived distinctions between nature or culture as the need arises, or to reconcile them. [Because men use what is known to them to make sense of what is not, a deep inquisitiveness about the relationship between electricity and the human body was part of the process of becoming socially acquainted with that novel and mysterious force in the late nineteenth century.] And though electricity might be discussed either as an extension of nature or of the body, or as something opposed to and outside them, it was defined in any case inescapably with reference to them.

The object of this chapter is to explore the relations between these constructed points—nature, the body, and electricity—in the imaginative worlds of popular and expert culture, and to examine the body as a communications medium, that is, as a mode for conveying information about electricity, and as a symbolic focus for hopes and anxieties entertained by experts and laymen about its significance. Though these two communities were organized by broadly different modes for sharing, discussing, and verifying information about electricity, both raised fundamental questions about it by exploring its relation to the body. Would electricity enhance and preserve life, and would it bring culture into greater harmony with nature? Or would it bring death and destruction, and further estrange what some regarded as a deteriorating relationship between culture and nature?

The Authority of Bodily Experience

Communities are conveniently defined by who talks to whom, what modes of discourse are acceptable, and what topics are discussed. As the organizing center of speech, gesture, and sense perception, the body was the principal cognitive instrument of laymen unskilled in the critical apparatus of literate experts. Its appeal lay in its apparent intimacy with nature, which made it seem particularly reliable. To laymen who embraced the body as a source of knowledge, nature was marvelous, with a marvelousness that did not have to be explained but was accepted as a gift. In popular science, or science as laymen understood it, the body was a reference point against which the whole world could be measured to make it comprehensible. When the *Electrician* declared that popular science appealed to the senses, it referred to a way of knowing quite unlike the controlled observational posture of empirical investigation.[1]

Popular electrical literature of the late nineteenth century attempted to anchor knowledge of electricity in sense perception and in nature experienced as a personal ally or enemy, a partner in a special and even magical oral-gestural dialogue. Men and nature were thought to be intimately in touch, and nature was thought to respond directly to human action and desire. The electrical press told many stories of ignorant bumpkins "wandering" into telegraph offices, as though vaguely aware of their operations, to "see it go"—to be convinced of electrical reality by concrete, immediate, sensible demonstration, rather than by abstract generalization. In such stories, the naïve inquirer might first attempt to determine if he were being trifled with, since electrical communication was not corporeally manifested in the expected way. While the mechanism of the telegraph was not transparent to immediate observation and required an abstract theory of the sort that scientists understood to supply the explanation of its operation, the wanderer nevertheless had seen it work with his own eyes. "Even then," went a typical story in this genre, "he could hardly be convinced, but he concluded that 'it was the durndest thing he ever saw.' "[2]

Expert culture, as we have seen in a preceding chapter, created its familiar electrical world out of formal theories and other print-based techniques of disembodied reasoning with specialized literate formulas and procedures. In scientific and technical literature, expert authority rejected immediate sensory judgment, or direct experience of nature, as naïve empiricism. In expert culture, nature was not a partner but a phenomenon for study, an object of mastery and conquest, something apart from man and understood through a screen of studiedly abstract models and theories rather than directly, that is, in dialogue. Amos Dolbear, an inventor in telephony, professor at Tufts University, and well-known scientific popularizer, dealt with the "all-embracing mystery of electricity" in a typically expert way in *Popular Science Monthly*. Though conceding that the nature of electricity still "befogged" even scientists, he also declared that the same scientists knew "pretty thoroughly what to expect from it" not by reason of vivid, one-time sensible proofs of the kind demanded by wandering *naïfs*, but because "it is as quantitatively related to mechanical and thermal and luminous phenomena as they are to each other."[3] The task remaining to science was to state the nature of electricity "in terms common to other forms of phenomena." To achieve the most widely generalizable abstraction was the goal of scientific knowledge. Like the scientist, the wanderer also hoped to establish the status of electricity by comparing it to the already familiar, but this was an implicit and unstated, rather than explicit, epistemological requirement; moreover, what counted as familiar to experts and laymen were different things. The layman's theoretical requirements were concealed in his sense of what was unalterably tangible. The irrefutably palpable circumstance was a good deal closer to the wanderer's sense of truth made firm. The scientist's faith in empirical demonstration was distanced from sense experience by layers of theoretical consideration that doubted the trustworthiness of appearance *a priori*.

Morally formulated divisions between oral-gestural and written modes are long-standing in Western culture. They run as deep as the philosophically constructed polarity between body and mind in behalf of which they are now and again summoned to do battle.[4] At the same time, practices associated with both modes, and with the rationales that justify them, are deeply ingrained and widely accepted. By the late nineteenth century, popular education, mechanized printing, and cheap paper had created a mass reading public accustomed to the habits of print, even if the form these habits took did not always satisfy the guardians of high culture. In parallel fashion, the community of science, self-defined by its practice of specialized literacies, was dependent on modes of apprentice training and demonstration that very nearly constituted a mimetic oral craft tradition. The association of popular culture with orality and expert culture with print literacy, in which the body is at greater remove from the phenomenon under study, was not, therefore, an absolute division, but one of relative emphasis. Because of the pragmatic command by both communities of a range of modal styles, both scientists and laymen on occasion embraced the logic of whichever tradition they were most skeptical of. At other times, the differences in the world each tradition accepted as real created tension between an oral tradition of bodily immediacy and the sifted abstractions of a skilled literacy.

To do battle with nature, experts were armed with special technical equations, observational techniques, goals for control, and formal conventions for constructing the knowledge they extracted from it. These conventions protected experts from unwelcome association with popular oral-gestural epistemologies. Experts were not, however, satisfied to talk only to themselves. Not only did their periodicals spend a great deal of time monitoring popular opinion on the subject of themselves and electricity; when they sought to present themselves to the larger world, their talk was laced with appeals to religion and magic, modes of experience vastly admired in oral-gestural culture. On these occasions, technical terms acquired priestly connotations, which experts probably were not sorry to see them acquire. And though laymen, or nonspecialists with popular views of electricity, clung to many of the oral-gestural beliefs and conventions that had always served them, they also wished to be part of and to count in the magical world experts dangled before them, and so they appropriated whatever notions of electrical terminology and expert procedures seemed serviceable. A single example may illustrate. "Fear," wrote one Dr. Cunningham in 1834, identified only by an expert title that conferred more than demonstrated literate authority, "is due to escape of electricity from the body, and joy to its entrance."[5] The same commentator held that electricity was responsible for the kinky hair of Negroes, the movement of the planets, and large feet among the inhabitants of the Northern Hemisphere. These "facts" placed bodily phenomena at the center of intellectual concern about electricity, and identified this set of propositions as a popular epistemology in which all things were connected, nothing could be accidental, and nature was powerful. While this epistemology was also recognizably scientific, connections among things in popular science were magical and direct; in expert science, they were theoretical, hierarchical, and provisional. Popular science respected the active power of nature, as expert science, which sought to organize and subdue it, could not. But experts gave credit to nature for having the final say about the worth of their theories. Dr. Cunningham's claims were given a scientific cast by their application to remote things (planets) as well as proximate bodily ones (large feet). As in science, a parsimonious explanation was offered for complex events, though the explicit connections required for a truly scientific explanation were absent, and not only absent but resisted, since connections in scientific explanations were often as mysterious to laymen as if they in fact were magic.

Nature as an Object of Expert Conquest

The nature behind electrical phenomena scientifically understood was complex and abstract, requiring training in abstruse vocabularies and apprenticed indenture to puzzling ideas. "We know little as yet concerning the mighty agency we call electricity," conceded William

Crookes in an article on the "possibilities" of electricity in *Fortnightly Review* in 1892, while presenting an impressive collection of arcane theories as to its character. "The only way to tackle the difficulty is to persevere in experiment and observation."[6] A single path to electrical knowledge required the disciplined denial of bodily perception much beloved by scientists, and was justified by its object, the conquest of natural vicissitude and the annihilation of interval in both action and communication across space and time.*

New modes of communication that vaulted across these boundaries signified human triumph over nature, increasing with each scientific advance. These triumphs made scientific investigators and technical experts creators in their own eyes of a new millennium, separated from a past in which men had possessed neither mastery of nature nor the enviable understanding of it that abstract knowledge had given them. Nature was the base line from which human civilization had emerged by progressively subjugating the natural. It offered the strongest possible contrast against which to measure and evaluate technological achievement with pride or, if one were so disposed, with alarm.

In expert epistemology, nature was messy. Technology was the great orderer. "The real calamity in a thunderstorm," explained William Crookes, commenting on natural unharnessed electricity, "is not that the lightning may kill a man or a cow, or set barns or stacks on fire. The real calamity consists in the weather being upset." The practical electrician should aim at "nothing less than the control of the weather" for the sake of agricultural productivity. Practically speaking, Crookes did not wish "to reduce our rainfall in quantity, but to concentrate it on a smaller number of days, so as to be freed from a perennial drizzle."[8] What he called "amending the ways of Nature" justified an expert, or adversarial, relationship to it.

What was messy was dangerous. In a speech to electrical engineers in 1890, Stevens Institute president Henry Morton suggested that

---

*The importance of bodily perception was often acknowledged by experts. Speculating on the possible existence of sentient beings with different sense organs than those of humans, Crookes wondered if such beings might have eyes sensitive to special vibrations of "electrical and magnetic" phenomena which would offer them "a different world from our own. . . . Glass and crystal [to them] would be among the most opaque of bodies. Metals would be more or less transparent, and a telegraph wire through the air would look like a long narrow hole drilled through an impervious solid body. A dynamo in active work would resemble a conflagration, whilst a permanent magnet would realise the dream of medieval mystics and become an everlasting lamp with no expenditure of energy or consumption of fuel."[7] Released from the strict bodily discipline of human science, the world of phenomena would be beautiful indeed.

the future of civilization hinged on the conquest of natural electricity. "Intelligently managed and controlled, the most powerful . . . agencies become the most efficient protectors and servants of man, and . . . aid him in his mission of subduing and utilizing nature," Morton declared. Without such servants, men would be "reduced to the lowest condition of savagery," helpless against "blind" nature.[9] Nature was the abyss ready to swallow culture. The threat of retribution by nature against technological order was always subtly present, as in a *New York Tribune* account of the reliability and safety of telephone service in the midst of an alarming winter storm:

> "I was stopping at a country house on an island near Stamford. It was blowing a furious gale of sleet and snow. The water was dashing madly against the rocks and the great trees about the house were swaying in the blast. All nature seemed to be in the wildest commotion, but the wires held fast, and when I rang up a friend in New York and his quiet voice came to me all the way through the wild night without a change in its tone it did seem almost marvelous."[10]

What was chaos could not assume the character of a dialogue with man, but was an unruly, unreliable, deadly force to be subdued. In such circumstances, the message from man to nature was not sent with any expectation of reply, but forcibly imposed. The desire to coerce nature pushed the expert off his peg of lofty disinterest and suggested the deeper emotional pull of the body even in expert science. The *Pall Mall Gazette* reported that the inventor Monsieur Rauspach had conducted experiments with an electrically charged prod on three lions, a boa constrictor, and an elephant, several of nature's most exotically dangerous creatures. Readers were treated to full details of how the animals were made to submit. The proud lions "were seized with trembling and growled fitfully." The twenty-foot boa constrictor "became at once paralyzed and remained motionless for six hours afterward. When he recovered he showed signs of numbness for three whole days." Touched merely on the tip of his trunk, the elephant "set up a series of wild cries, and became so enraged that the tamer feared the brute would break its heavy iron chain."[11] *Popular Science News* reported in 1897 that electricity had been used to "conquer" a recalcitrant horse. "In one case a very high-spirited and valuable animal, but extremely vicious and balky, was cured in one hour with the aid of a three-volt dry battery."[12]

Richard L. Garner, a naturalist interested in the habits and "speech" of the great anthropoid apes in the African wild, designed a special electrified steel cage and lived in it many months, keeping nature at a technologically remote distance. His "fortress in the jungle" was a depot for supplies and a "place of safety from the wild beasts that prowl through the forests at night." It contained a phonograph, photographic instruments, several telephones (connected to what was unclear), and an electric battery good for three hundred hours. Prior to an expedition to the French Congo in 1892, Garner explained to readers of the *North American Review* that by means of a small switchboard

> I shall be able to fire my flash light at night or to snap my kodak in the daytime, and to operate my telephones if necessary. In case of danger or unexpected attack, by the use of my switch-board and by means of a Stilltion coil, I can charge the entire cage with electricity, developing an alternating current of about 300 volts. In leaving my cage with its contents for any length of time I shall simply charge it in this manner with electricity, in order that in my absence my meddlesome neighbors may be induced to let it alone.[13]

Another experiment described "a large, ugly spider that had been feasting on flies for two months," hopelessly befuddled by a tuning fork set to vibrate against its web, which was conceived by the experimenter as a kind of primitive telephone network. Expecting a buzzing fly, the investigating spider could not decode the source or content of the odd message. The strong god-playing element in this story paid homage to belief in the superiority of technologized intelligence. Man with his tools knew better than nature; he had foiled and humbled it, laughed at its ignorance, and made it run to do his bidding.[14] In each of these expert tales, the underlying fabulous and mythic encounter between nature and the magic wand of human invention—a melodramatic, magical confrontation full of the stuff of fairy tales—was disguised in the vocabulary of scientific knowledge and achievement, which doubtless made it convincing to its audience. Monsieur Rauspach's experiments, for example, had been reported to the Academy of Sciences.

While technology held the line of civilization and kept savage nature at bay, no reciprocal moral sense constrained men from using technology against nature. Nature damaged or threatened by electricity was evidence of human triumph over instinctual forces. It was said that electric lights on the Capitol Building presented an untidy appearance because of the

> billions of insects which have been drawn thither by the brilliancy of the electric lights . . . whose skeletons are either hanging on the walls,

held fast by a death grip, or are piled up in heaps all over the recesses of the roof. May flies, beetles, crickets, earwigs, dragonflies, grasshoppers, caddis flies, bees, wasps, ants, hornets, skippers, horned midgets, gnats, mosquitoes, and every species of insectoria known to the surrounding swamps and woods of the district.[15]

Nature encroaching on civilization was sacrificed at civilization's altar, repulsed by a deadly weapon at the center of civilized accomplishment. If nature repulsed or inconvenienced mankind—if centipedes, "those 'horrid thousand-leggers,' " terrified the ladies—it could be gleefully noted that electric lights rapidly dispatched "the ugly little bodies," and that with this application of the miracle fluid "there is, apparently, no end to the uses of electricity."[16]

In the late 1890s, a series in the *American Electrician* detailed uses of electricity by power station technicians for "sport." Contributors described their amusement when streams of electrified water were turned on stray dogs, occasionally electrocuting them, when wires were wrapped around chunks of meat to bait serenading cats that were shocked into entertaining leaps and somersaults, and when dynamite was electrically detonated to destroy an unluckily located nest of yellow jackets.[17]

But what would it mean if the weapons of control turned against the controllers? At the dangerous edge of civilized behavior, electricity could be an ominous symbol of projected anxieties, a signifier not of order but of instincts indulged at peril. With President Porfirio Díaz and members of his cabinet in attendance, the first use of electric lights at the bullfight in Mexico City in 1887 threatened to pitch the excitement of that volatile drama too high. Ten arc lights mounted for the occasion seemed to make the bulls wilder than usual, and "the gaudy uniforms of the matadors fairly blazed."[18] The same year, extensive advertising promised that the annual Bull Circus at Nîmes would be conducted beneath electric lights. When the lights failed at the beginning of the performance, the enraged audience rioted and made a bonfire of the circus fittings before troops managed to restore order. On such occasions, electric lights seemed to lend power to instinctual forces straining the fabric of civilization.[19] Another reversal of the expert-conquest theme was stories in which man had so mastered nature that the line between nature and man's artful transformation of it was no longer clear, and nature seemed more sensibly depicted in the image of human creation. In the spring of 1887, for example, it was said that the appearance of Venus in the early evening sky confused observers who inquired of the *New York Sun* whether the object in question was not an electric light sent up in a balloon by the ingenious Mr. Edison.[20]

Nature in Dialogue

If experts believed that electricity could be known only by experiment and observation—procedures not for speaking to nature, but for capturing her secrets in order to force her into technological servitude—those committed to the epistemology of bodily authority believed that nature revealed herself to those who waited for her to speak in her own language. Electricity, X rays, and wireless were not the exclusive property of science, but had been there all along for those with eyes to see their manifestations. In 1896 the *Buffalo Courier* reported that Dr. John T. Pitkin of that city believed that X rays were frequently exhibited in nature. The doctor was quoted as having "seen through a tree during an electrical disturbance."[21]

Folk wisdom affirmed natural miracles that science was expected to witness and verify rather than explain, since explanations suggested that things were other than they seemed, and folk wisdom accepted events as given. Scientific terminology lent an air of up-to-date credibility to remarkable things seen with one's own eyes. Readers of the *Telegraphic Journal* of London learned the "well-authenticated fact that certain flowers such as marigolds, sunflowers, and poppies have been seen at rare moments to emit little flashes of light," especially during the hottest months, after sunset or before sunrise, and during drought—a phenomenon solemnly attributed to the inductive effect of atmospheric electricity.[22] *Popular Science News* reported that Maurice Depres, an electrical engineer in Cordova, Spain, had witnessed a shower of electrified rain on a warm and windless day when the setting sun had been overcast with dense clouds. Soon after dark, as lightning flashed, "great drops of rain fell, which crackled faintly on touching the ground. From each of them sparks darted towards the walls, trees, and soil they fell upon" for a brief time.[23] In Hungary an American "professor" had succeeded, he said, in bottling fresh lightning for local farmers to put on their fields to stimulate rain whenever their crops needed a drink.[24]

A view that natural manifestations were part of a dialogue between man and the world saw nature's retreat before technology as a threatening development, and not as a positive sign of man's mastery of chaos. In rural Iowa, where farming required a partnership with nature, a local newspaper wondered what would happen to the cows each night when the electric light was introduced. Would they know how to sleep? "Is there a town over the broad earth where cows run loose under electric lights?" Pondering the unseemliness of this hybrid of nature and technology, it added, "A town with electric lights and cows running loose in it would be a spectacle for gods and men, resembling a savage clothed in a silk hat."[25] The *Milwaukee Sentinel* complained that electric lights constituted "a very bad and wholly unnecessary imitation" of daylight, as though "Nature made a very grave mistake in instituting darkness, and the arts of men are engaged in efforts to correct Nature's blunder. The planets are too far off to afford any useful light, the stars are useless, and the moon is too irregular in its habits as a luminary."[26]

Discomfort with the menace of electrical technology was elsewhere manifested in apocalyptic theories of disaster. One of the most popular was that excess charge accumulating in the world posed a growing danger to man and nature. "What would that class of theorists do with electricity without that poor, bamboozled and bedraggled word 'charged?'" wondered the *Electrical Review* in 1886. "It is made to do duty on every occasion, when there is uncertainty, perspecuity [sic] or indefiniteness. The ground, the wire, the machine, the air, the clouds, are constantly 'charged.'"[27] Suggestions were put forward that the amount of lightning in the air was increasing as a direct consequence of the spread of telegraph, telephone, and electric light wires across the country. The *New York World* editoralized:

> The proposition is advanced that pretty much everything that will hold electricity is becoming more or less charged. Fears that have not yet assumed a definite expression are entertained by many observing people to the effect that too much of the subtle fluid is being manufactured and kept in store to be consistent with the public safety. It is thought that much leakage is involved and that the earth, especially in the case of large cities, and the houses are being more or less saturated with it. It is time to call a halt before this thing goes any farther. With telephone, telegraph, and electric light plants on the increase, and the electric motor still to come, the situation is serious and demands prompt attention.[28]

A slightly different hypothesis elaborated by a certain E. Miller of Kingston, Missouri, also gives the flavor of popular misgivings:

> The iron rails as they lay on the ties are great conductors of electricity, and so are the wire fences. These are the disturbing elements. This unusual electricity is collected by nature in large and positive bodies, we

know not where. When the rain cloud comes up, these bodies of electricity are attracted to them in such ponderous masses that the clouds cannot neutralize them, and sometimes change the course of the clouds, and then we have a cyclone; or sometimes they take the same course as the wind, and then we have a hurricane or straight wind, as they are called. But the disturbing element is all the same.[29]

Miller advised that requiring railroads to put down ground rocks six feet in length at two- or three-hundred-foot intervals, and requiring farmers to put down ground wires at the same intervals along their fences, would eliminate tornadoes in the state of Missouri. From Bishop Turner of the African Methodist Church of Georgia, Kentucky, and Tennessee came a related warning concerning the spread of electric light. "While admiring the invention of the white man in controlling electricity," *Electrical World* reported,

> he claims that the subjection of God's agent is carried too far in making it light the world. He predicts that the unbalancing of the air currents which electrics are causing will in a few years, if they increase in numbers as fast as in the past five years, cause whole cities to be blown away at a time, and floods unlike any save Noah's. All the floods, hurricanes, cyclones and other atmospheric disturbances taking place in the heavens and upon earth are due to the work of electric lighting companies, says the reverend bishop.[30]

Not even those "best able to express an opinion," scientific experts, had a proper explanation for tornado activity, the *World* harrumphed in scorn. The bishop must hold gas stock or be under contract to the gas interests of his diocese, speculated the editors, who could only thus explain one who did not admire the white man's invention enough.

The electrical press was equally disdainful of a physician's estimate, obnoxious to them for its quasi-scientific statistics, that the peril to human health from electricity was "three to five fold greater than it was fifty years ago." Disparaging and lumping together Negroes, literature, and superstitious magic at a stroke, the press claimed this theory would "do honor to a colored revivalist, or one of the witches of Macbeth."[31]

Expert ridicule rarely deterred cosmic tragedians. In 1888 the Reverend A. C. Johnson prophesied that the very days of the earth were numbered by man's extravagant production of the lightning. "In just 32 years from now," reported one account of this prophecy, "the electricity stored in the earth will come in contact with the heated matter inside and blow the whole world up." The spokesorgans of expertise sarcastically suggested avoiding this little explosion by tapping the reserve of stored electricity for telegraph and streetcar service.[32]

Other observers couched their concerns in more scientifically imitative, though not necessarily accurate, terminology. A British observer accepted an upset in the equilibrium between electricity and the rest of nature as a plausible explanation for a stretch of abnormally wet weather in 1892. "Electricity," he explained, "is a *palpable* substance, universal in extent, in the earth and in the atmosphere, and upon it apparently depends the fertility of the earth and the vitality of the atmosphere."[33] Unfortunately, the commentator continued, mankind's constant manipulation of electricity endangered earth and air alike:

> We force the electricity out of the earth by both chemical and mechanical power, and having used it according to our will for illumination, motive power, or other purposes, *throw it into the atmosphere,* which, being already replete, cannot absorb it. It will therefore be naturally attracted to the oxygen and hydrogen gas, and united with them turn them into water, and thus descend again to the earth, after having at once weakened the fertility of the earth and the vitality of the air.

Recurring in all these expressions of popular concern was the fear that man was throwing nature wildly out of balance in his manufacture of electricity, and that nature would sooner or later redress that balance. If man misused the cosmic order, the cosmos would take its revenge. Nature might be pacified if man were to revere it as an equal partner instead of disdaining it as a slave to man-made science. The alternative possibility, that tragedy loomed over clumsy human efforts to prize loose the secrets of the universe, weighed heavily on the popular imagination.

At the opposite end of the scale from predictions of cosmic disaster, but with a still more powerful hold on popular thought, were reports of electrical accidents daily occurring in plants and in the streets. In the public perception, electrical accidents were an increasing risk of urban life. When these reports were picked up by the technical press, experts commented on the errors and exaggerations that often accompanied them in order to argue that the ignorance they perpetrated was a more serious social danger than any physical threat from electricity.[34] This approach also strengthened the case experts never tired of making for greater expert control over electricity, which the Kentucky Court of Appeals described as "the most powerful and dangerous element known to science" in an accident-case ruling. Electricity, the court added, was "a force which no one except experts can understand."[35]

What made accidents more fearfully compelling to popular audiences than any expert reassurance or irritation to the contrary were vivid and grotesquely detailed descriptions of what these accidents did to the body, that most informative of instruments for conveying to laymen what electricity was really like. "How a 2,500 Volt Shock Feels" was a typical title in *Popular Science News,* for the experience of an electrical engineer who had lived to tell the tale.[36] A Texas lineman who survived a 500-volt shock described it in memorable terms:

> Just as the shock came, it seemed to me like a blinding flash of lightning had come and its lurid flame had struck and remained in my eyes. I then felt myself reaching upward and as though I was going to fly, this, I suppose, must have been the sensation caused by the contraction of my muscles. Then I felt as though I was soaring away, just as one feels when put under the influence of ether or chloroform. Then all was blank and I knew no more until I felt a pricking sensation in all of my members. I then felt a strong tase of brimstone in my mouth, just as I am told that those who are struck by lightning experience always.[37]

Gruesome and serious injuries were reported from thunderstorms, from handling electrical materials, and from pranks upon or by the unwary.[38] *Western Electrician* reported how a lightning storm had electrified a swing bridge in Chicago: "A driver urged his horse out upon the bridge in spite of the blue flames that were playing along the iron rods. The animal was hardly upon the structure before the electricity leaped up through the iron calks of its shoes and it went down in a heap, stone dead."[39] Unfathomable mysteries caused other accidents. "It is said that in the dry atmosphere of the elevated plateaus of the Sierra Nevada and Rocky Mountains the human body becomes highly charged with electricity; and two serious accidental explosions which took place recently are attributed to this cause," reported the *Telegraphic Journal* of London.[40] "These are the days," commented the *Electrical Review,* "when numerous more or less excited individuals see balls of fire the size of watermelons running along telephone and telegraph wires and exploding with a loud noise and sulphuric smell."[41]

The Current of Life, the Paradise of the Body

As easily as it mobilized social anxiety, the inscrutable ether inspired visions of Edenic abundance, a paradise of the body. Scientific, lay, and religious observers were struck by the thought that because of its greater versatility and efficiency than steam, electricity might be used to indulge every frivolous whim as easily as to fill every basic need. Popular attitudes were as suspicious of the body as a source of pleasure as science was suspicious of the body as a source of knowledge. This created moral perplexities. If humanity in an electrical age would not want for food or shelter, then electricity must be a gift of God's grace. But what was the moral status of the "electric cocktail," described in 1877?

> "Electrical cocktail" is the latest. A flexible lead from the electrolier ends in a platinum curl. A trifle of sugar is added to the liquid, the platinum curl lowered into it and current turned on to make the curl red hot. A small amount of the alcohol and sugar is decomposed, i.e., carbonized, and the resulting burnt sugar is said to be very delicate. It promises to be a fashionable winter beverage, and can be made cold or hot.[42]

Or what of electric ornament? The twentieth triennial exhibition of the Massachusetts Charitable Mechanic Association in Boston in 1898 featured candelabra made by tipping "the branches of the antlers of one of the most perfect bucks' heads that the writer has ever seen" with incandescent lights manufactured by the Buckeye Electric Company of Cleveland.[43] "Flash" jewelry from Paris exhibited in New York in 1884 included hatpins and brooches studded with tiny, glittering electric lights mounted like jewels. One expert journal strained to discover some utilitarian function for this extravagance of electrical ingenuity:

> The practical use we see in these jewels is that, in returning home late at night, they afford a ready means of brilliant illumination, which would aid in the finding of a lost object, or one's way, and also the way to the key-hole, etc. It is said that the walking-stick, provided with a large diamond, affords sufficient light to read a newspaper by. If set with say a white gem on one side and a red one on the other, it may be used for signalling to a distance, while the switch would enable a communication to be carried on by means of the Morse alphabet.[44]

The electric pushbutton, another luxury artifact, symbolized a streamlined consumer electricity capable of delivering instant gratification. "I press the Dictionary button, and the Dictionary tells me whatever I want to know," explained a John Kendrick Bangs character in a fiction story for children.[45] Real pushbuttons were marketed more prosaically. Western Electric Company of Chicago peddled a modest dining-room pushbutton in 1890. Its claim to versatility was that it could be attached to the edge of a table, a side panel, or a chair, to call the servants.[46]

For elites, pushbuttons often symbolized popular desires for dangerously superficial pleasures; for laymen, pushbuttons often signified a world in which decisions about technology were taken beyond the control of ordinary people. The five-penny American magazine *Yellow Kid* featured an 1897 cartoon of a gnomish, slender man with a frown on his small worried face beneath an overdeveloped, egg-shaped, bespectacled, bald cranium. Leaning on a cane to support his fragile weight, the little man impatiently pushed a button with one disproportionately swollen finger. "An average man of the drear future," read the caption beneath, "when everything is done by pressing a button."[47] The socially and professionally conservative British journal the *Electrician* admonished: "It seems to us that we are getting perilously near the ideal of the modern Utopian when life is to consist of sitting in armchairs, and pressing a button. It is not a desirable prospect; we shall have no wants, no money, no ambition, no youth, no vices, no individuality."[48]

Traditional utopias of perfectly deployed technology knew no shortages of basic necessities of food, clothing, and shelter. Neither were there excessive desires in these utopias, no cultivation of acquisition, or pursuit of novelty for its own sake. Provision for need in utopia should not make gluttons or sybarites of its citizens. In an electrical age, however, men might want what was bad for them and be able to have it. The paradise of fulfillment might lead them to greater avarice, and not to contentment and self-restraint. Electric frivolity suggested the possibility of bodily self-indulgence on a grand scale.

Perhaps this was why religious authors viewed the extension of the Gospel to barbarous savages with enthusiasm while the electric distribution of the Word to their own parishioners filled them with unease. In Tunbridge Wells, England, efforts to enlarge the flock of a Congregational church beyond the sanctuary to sixteen telephone subscribers, including physicians, druggists, clerks, "an invalid lady who has been obtaining consolation from the telephone for several months," and "some lazy club men," were viewed with deep suspicion by local churchmen. Only the invalid lady received the benefit of the doubt; the rest were suspected of harboring the "spirit of experiment" or a desire for "entertainment," or attempting to "arrange the length of the sermon *à discretion.*" If the physicians could spare the time from their patients, what kept them from being present "in bodily form" in church? The critics concluded:

Whether it is worth while to turn the telephone into a preacher for [invalids], while serving as a medium for luxurious ease and novelty, and a promoter of church absenteeism, is a question which each reader may answer for himself. . . . We may seriously question . . . if that ingenious instrument will hasten by one second the dawn of the world's Millennium Day.[49]

Beyond this peril to piety, what purpose would religion serve if the burdens of life for which it was a solace ceased to exist? The belief that the pain and struggle of existence would be rewarded only in eternal life, and that poverty was blessed, was not first challenged in the nineteenth century by electricity. But the possibility that electric technology would turn extravagant luxuries into the furnishings of daily life made its challenge to traditional religious values more insistent. Electricity raised the specter of earthly affluence as it did not seem to have been raised before.

Electrical theology sketched two broad apologetics for the prospect of unlimited prosperity. In the first, electricity was a gift of God, created for His purposes and conferring obligations on men to further those purposes by using it wisely. Lack of respect for the Deity's property, the sin of the original fire-stealer Prometheus, and the sin as well of Adam and Eve, was the cause of a number of evils in the world, among the most serious of which was loss of faith. An apocryphal story circulated about Michael Faraday, the discoverer of electromagnetic induction. It was said that Faraday had been "put away" from the small Sandemanian congregation attended by generations of his family because his scientific researches had unsettled his belief. The congregation prayed for their lost sheep's spiritual return, and their prayers were answered. One day the man honored by "all the world" returned and confessed with tears in his eyes. The Sandemanians "considered it a terrible thing for a good man to devote himself to such doubtful subjects as electricity, instead of reading the Bible and being satisfied with things as they were."[50]

To keep men mindful of God's goodness in creating electricity, the Bishop of Aix, France, consecrated a new electric light plant to God's work in 1896. He observed with regret that authors on electricity frequently overlooked its supreme Author. Electricity had its foreordained part to play in God's plan to establish His spiritual kingdom at the end of time:

And man has appropriated this terrible fluid; he has imprisoned it in his apparatus; he has made a circle of wires around the globe; he has placed his wires at the bottom of the sea, which has thus been tamed to his service one time more. He has said to the lightning: "Behold the road thou shalt follow, thou shalt go where I command, and stop when I wish; thou shalt carry my thoughts to the farthest islands. I will condemn thee to the most prosaic uses; thou shalt light our streets, our houses, our shops, our churches; thou shalt serve us for rapid coursers on our roads and on the seas; and we are not yet arrived at the limits of the benefits of thy power, which has no equal in the masterpieces of creation.[51]

The Reverend L. Osler of Providence, Rhode Island, preached "in words that burn" that God had reserved the discovery of practical uses for steam and electricity to "these closing days of this world's history, when the King's business would require haste."[52]

A different religious interpretation of electricity avoided the moral dilemmas of electrical prosperity by proclaiming all electrical progress a spiritual triumph. Electrical science was treated as an extension of religious revelation. This approach did not subordinate the physical universe to its creator, but identified *all* forces in the universe, including electricity, with God Himself. By fusing scientific and religious vocabularies, electrical pantheism also skirted the conflict in which superior truth was claimed separately for faith and science by their respective devotees. A characteristic metaphor of reconciliation in this uneasy equilibrium was the statement of one amateur student of electricity that "the Deity is the omnipotent magnet."[53] In a speech to the New York Electric Club, electrical entrepreneur Erastus Wiman put forward an appealing synthesis of faith and science:

Once there came to this world a Saviour, whose life was a strange admixture of good to the body and good to the soul. When He left this world, that which was good for the soul He left behind; and the gospel of kindliness, the spirit of doing unto others as you would be done by, and all the great and glorious influences of Christianity were inherited by poor humanity that sadly needed them. It was supposed, however, that He had gone and taken with Him the power of curing, and that no longer would the miracle be performed, no longer would the touch cure, and that the influence which on earth He used was lost to mankind forever. But it may not be so. That power we speak of tonight, that marvellous power which is so subtle, which is so mysterious, is equally miraculous. It is available for us all, is with us still, a revelation to us. We should thank God that we live in an age and in a time when, with these electrical minds that are before me now, we are daily discovering

new uses for that power, and among them not the least the power to
mitigate human suffering and prolong life.[54]

Even the Bishop of Aix's notion of electricity as God's gift struck
a pantheistic chord by describing electricity as immanent in all creation: "It is everywhere, and there is no being which it does not lurk
in; it is in my hand, in my gestures, in my voice, upon my tongue, in
my entire system, and in this sheet of paper. It is as the invisible soul
of the material world."[55] This description by a member of the clergy
was not unlike that of Heinrich Hertz, one of the most respected of
scientific authorities. In an 1889 address to a Heidelberg medical congress, Hertz prophesied that electricity would soon be understood as
a pervasive cosmic phenomenon:

> We shall see henceforth electricity in a thousand circumstances where
> we did not suspect it before; each flame, each luminous atom becomes
> an electric phenomenon. Even when a body does not give off any light,
> provided it radiates heat, it is the source of electric actions. The
> domain of electricity thus pervades all nature—it pervades ourselves;
> in fact, is not the eye an electric organ?[56]

The power of electricity was sometimes compared to the power
of the sun, whose limitless capacity to sustain life was an old and
potent motif. Solar energy and the wireless transmission of electric
power were both utopian dreams of the late nineteenth century. Electrical entrepreneur Wiman predicted that electricity would one day be
gathered directly from the sun, with no intervening devices.[57] A nearly
mystical vision of wireless power as a life force equivalent to the sun
came from Amos Dolbear, who believed that

> One may . . . see how life might be maintained without foods of any
> kind in an organism adapted to absorb energy from space about it as a
> piece of matter gets warm in the sunshine, for it is beyond peradventure
> that all space within our ken is saturated with energy, that electrical
> energy is present in every nook and corner of creation and that there is
> no point in space where an abundance is not to be had if the demands
> of life should depend on it.[58]

With the intensity of personal acquaintance with revealed truth,
some observers were certain that electricity contained the divine power
to bestow life. If other elusive powers had been harnessed by the labor
of intellect, why not this Olympian craft, which seemed also to lie
within the grasp of unfolding scientific knowledge? Immortality, like

the telephone, could be invented. And if electricity could not yet stave off death forever, at least it could prolong life or, as a very last resort, rescue those in a state of suspended animation who were mistaken for dead. A widely reported invention for this purpose was the electrical grave-annunciator. In 1891 William H. White of Topeka patented a device that could be attached to the hand of any doubtful corpse to signal electrically to those aboveground in the event of belated revival. In another system, the coffin was filled with compressed air, so that the slightest movement would activate an electric alarm button.[59] More elaborate arrangements were possible with the advent of wireless telegraphy. *Answers* reported in 1899 that a California millionare "is to be interred, according to his express instructions, in an open coffin; and above his head, within easy reach, is the tiny sensitised vial which, should he awake, will flash outwards and upwards from the silence and darkness of his mausoleum the news of his resurrection."[60]

In sum, efforts to ally electricity with technology for the purpose of extending technological control spurred attempts in other quarters to reconcile electricity and nature. A common solution made the human essence electrical and one with the universe, and held that electricity was the life force immanent in the whole material and immaterial world. In this case, man could not separate himself from nature to rule it, because he was of it. Another solution was to identify the power of electricity with the sun, the great life-giving force beloved alike by science and myth, or to invest electricity with the secret of life, which might yet be discovered. This last view cleverly unified scientific and religious authority, always at odds over the status of electricity. But its secularization of traditional religious themes caused as much theological unease as its metaphysical leaps brought indignant scientific disclaimers.

The Electrically Transformed Body

An important set of themes about electricity and the body pondered the body physically, medically, or culturally transformed by efforts to use electricity as a healing agent. In popular superstitions surrounding persons who seemed possessed of special electrical powers, and in discussions of electrocution and fantasies of electrical warfare, electrical ornament made the body it embellished an object of cultural focus and fascination, and a powerful medium for messages of civic, sexual, or class status.

Electrical Healing

Popular belief in the existence of a mysterious vital force at the root of life was receptive to the notion that this force was most likely magnetic or electrical, that experts would soon harness the powers of electricity to cure bodily ailments, and that particular diseases and forms of distress were traceable to electrical imbalances of one kind or another. A legitimizing step by experts was to create a mythic past and purpose for medical electricity, part of a historically continuous scientific enterprise to redeem the body. The first appearance in English of the word *electricity,* according to Park Benjamin, was in a 1650 translation of an earlier treatise on the curative effects of magnetism written by John Baptista van Helmont, a Flemish physician, chemist, and Rosicrucian.[61] Galvani's discovery a century and a half later of "animal electricity" (which he declared the basic life force after finding that electric current caused a frog's severed leg to kick as though alive) was frequently recalled as an early medical experiment.

Heroic figures were said to have performed early cures with electricity. The French revolutionary hero Marat had restored the sight of a blind man with electricity, according to one tale.[62] Somewhat less prophetically, during a severe yellow fever epidemic in Jacksonville, Florida, in 1888, a Kentucky physician suggested to the surgeon-general of the army that the "delicate and subtle" atmospheric poison of yellow fever might be dissipated by strong electric light. He recommended erecting a row of large army tents along the center of a Jacksonville street set on either side with electric lights "so intense as to repress the poison and stay the destroyer."[63] The electric light was also introduced into the body for medical purposes. "Who could ever have dreamed that the electrical current manufactured by the public lighting companies conveyed along the streets would be switched off on special wires to go into the very mouths of people," marveled the *Philadelphia Inquirer* in 1889.[64]

Reports on medical experiments with electricity were a commonplace of electrical literature. In 1889, a London physician reported that the growth of a cancerous tumor had been arrested by several months' treatment with electricity.[65] Successful treatments for diabetes, gout, rheumatism, and obesity with high-frequency currents, said to work by "augmenting organic combustion," were reported to the French Academy of Sciences in 1896.[66] St. Luke's Hospital in New York underwrote experiments to introduce electric current into the lungs in sufficient quantity and strength to kill tuberculosis bacilli.[67] From time to time researchers discussed whether drugs could be introduced into the body on the model of wireless transmission.[68] Belief in the healing benefits of electricity for humans was extended to other forms of life. Sir William Crookes reported that efforts to increase crop production by means of electrical stimulation not only seemed to give "increased vigour to the life of higher plants, but tend to paralyze the baneful activity of parasites, animal and vegetable."[69]

With the example of expert medical interest in electricity before it, popular culture accepted electricity as the active therapeutic agent in pills, soaps, teas, potions, lotions, apparel, and jewelry of all kinds. Quack nostrums boasted miraculous electrical properties, like the snakebite remedy promoted by Patrick Cunningham of Richmond, Kentucky: "When the snake's fangs strike this liquid in the human body, an electric current is generated, which drives the poison in the reptile's body through every blood vessel in its system, causing almost instant death."[70] When researchers at Vanderbilt University announced in 1896 that X rays caused hair loss, a French entrepreneur guaranteed by this means "to remove the mustaches and whiskers with which some French women are adorned."[71]

Customers of a London wigmaker could buy a special battery apparatus that, placed in the lining of a silk hat, diffused a current "all over the wearer's head" to cure nervous headaches and neuralgia, and to alleviate baldness.[72] Less specific in its claims was a sign in a shop-window in a small English country town: "Try our electric tea—good for the nerves."[73] Electricity "improved even the bath," reported *Science Siftings* in 1896:

> A battery has been patented consisting of a source of electrical energy placed inside a cake of toilet soap. The device is reputed to be intended for curative applications of electricity to the human body. The inventor says that his invention is based on the fact that the chemical decomposition of soap is such that when dissolved in water it produces a liquid having an exciting effect upon certain metallic electrodes placed in proximity to form an electrical battery. In other words, you soap yourself and get an electric shock.[74]

A special basement room in the Capitol Building in the late 1880s was fitted out for congressmen to indulge what the *Electrical Review* described as "quite 'the fad' nowadays" of taking their electricity:

> An electric apparatus has been fixed up in the engine-room in the basement and daily the members avail themselves of the opportunity to get freshened up. A board, with a tooth-piece of copper, is placed beneath

the great belt of the large engine-wheel, and the electricity thus generated is carried off by a wire attached to the board, which is long
enough to be grasped by one who sits in a chair near-by. The circuit is
completed by the person holding the wire grasping a small brass chain
attached to the railing around the engine's wheel. The system is thus
filled quietly with electricity. The members say it is splendid after they
have been out to receptions and suppers all night, or after they have
exhausted their brain power by speechmaking or listening. A great many
members take electricity, and some go to the basement of the Capitol
for it every day during the season.[75]

Virility, long associated with terms like *force* and *energy, strength*
and *vigor,* which also described electrical properties, was an area ripe
for electrical theorizing and therapeutic promise. A durable popular
whimsy that the secret of eternal youth must be connected to electricity
was transferred with special intensity to sexuality. The mysteries of
both sex and electricity, deepened by ignorance of the physiology of
one and the physics of the other, made these two natural allies in the
hands of skillful ad men.

The "Heidelberg Electric Belt" sold by Sears, Roebuck and Company promised to make a new man of its wearer in a month. To recruit
customers, Sears rented the entire available stock of half a million
"nervous debility" letters held by letter brokers. These were letters
responding to ads that offered remedies for various real or imaginary
male ailments. Each correspondent received a fifty-seven-page electric
belt catalogue.[76] The Harness Electric Belt and Suspender, the British
counterpart of the Heidelberg Electric Belt, was the object of a determined campaign of exposure by *Electrical Review, Science Siftings,*
and sister publications that hoped to make an example of it as a fraud.
The Medical Battery Company, manufacturers of the Harness Belt,
and its spokesman, Dr. Tibbets, countered with libel proceedings against
the *Electrical Review* in 1892, which were ultimately unsuccessful.[77]
Harness advertising was widely distributed through periodicals such as
Answers, the flagship periodical of the Harmsworth publishing empire,
a powerful channel to British mass readership. Typical of Harness advertising was the claim "It cannot fail to restore impaired vigour, and
speedily renew that vital energy, the loss of which is the first symptom
of decay."[78]

Electrical corsets were available for women, less to augment their
sexuality than to control it. "If one of these articles is pressed by a
lover's arm it at once emits a shriek like the whistle of a railway engine," one inventor promised. He claimed to have married off three

daughters "owing to the publicity" thrust by this device on their hapless suitors.[79] In keeping with Victorian conventions of feminine modesty, electrical advertising for the treatment of female troubles did not
dwell exclusively on sexual maladies, but promised to cure a variety
of ailments that suggested the delicate health attributed to genteel women
in this era. Much of this advertising had a religious cast. One ad that
appeared regularly in *Comfort,* a popular women's magazine, capital-
ized on the notion of wireless transmission as well. For the convenience and privacy of bedridden *Comfort* subscribers across the United
States, Professor S. A. Weltmer of Nevada, Missouri, self-proclaimed
"Wizard of the West," possessed the power of curing patients at a
distance. Following "in the path made by Him who was born at Beth-
lehem," Professor Weltmer achieved his cures with the Method of
Magnetic Healing, practiced "without drugs or the surgeon's knife."[80]
Constipation, indigestion, poor circulation, liver, stomach, and heart
trouble, eczema, female trouble, ulceration, and general debility had
all been cured by Weltmerism after every other medical route had been
exhausted.

Electrical phenomena were also regarded as a source of bodily
distress and disequilibrium. "As civilization advances," the *British
Medical Journal* reported in 1889, new kinds of diseases were produced by "novel agencies which are brought to bear on man's body
and mind."[81] One of these new conditions was "aural overpressure,"
to which even "strong-minded and able-bodied men" were susceptible
because of the "almost constant strain of the auditory apparatus" in
persons who used the telephone for "a considerable portion of each
working day." Sufferers from this malady, some of whose ears were
also irritated by the "constantly recurring sharp tinkle" of the bell,
experienced nervous excitability, buzzing in the ear, giddiness, and
neuralgic pains.

In 1890 it was reported that the telephone had driven a Cincinnati
citizen insane. The victim had always been "excessively nervous." The
constant ringing of a telephone in his office finally drove him mad.
His symptoms were peculiarly manifested. "He rushes into drug stores,
rings up the central and calls for his departed sweetheart."[82] He was
not alone. Mrs. Mary Winsgate of Illinois, refined, well dressed, and
possessed of a low voice, had given the following testimony to the
judge who committed her:

> "I must find Edison today; he's the telephone man, isn't he? He has
> stuck telephones all over my house and one on the gate, and they keep
> ringing all the time, day and night so nobody can rest or sleep. Some-

times I think it is the door-bell, and when I go to the door there is a little red-headed woman who slaps my face and runs away without saying anything. There is a black flag on my house right now. What is it there for? I pay taxes and nobody has the right to put a black flag on my house. We're all metalized—everything's metal in my house. I'm metal and so is my son, and there's a battery in the basement which has charged every one of us."[83]

A man in the Hudson County insane asylum in New York began to hear voices speaking to him. Upon hearing a real voice through the telephone, he became violently insane and smashed the instrument to pieces. Subsequently, he claimed to have a telephone in his stomach, through which someone periodically talked to him.[84] Other electrical media were accused of affecting the "nerves." Electric light was blamed for producing a special form of ophthalmia. Prince Bismarck was said to have succumbed to nervous debility from the telegraph, which condition precipitated his withdrawal from the Prussian ministry.[85]

Special Powers of the Body and the Body Inscribed

Just as the power of witchcraft was imputed to exceptional or marginal individuals in other cultures, in popular electrical culture superstitious fear and respect were accorded those whose bodies were said to be invested with special electrical powers. In 1886 the *Electrical Review* reported the story of eleven-year-old Willie Brough of Turlock, California, whose neighbors in the San Joaquin Valley believed he could set objects afire by gazing intently at them.[86] After five unexplained fires had broken out in Willie's schoolhouse one afternoon, the schoolmaster forbade his return and declared Willie a victim of supernatural agencies. Social pariahs, Willie's family moved to a remote cottage in the cottonwood timber across the valley. Investigation determined that Willie, "an extremely nervous boy . . . with a largely developed head," was actually "overcharged with electricity." Sparks appeared when he snapped his fingers, and Willie admitted seeing sparks fly around him in the darkness of his bedroom.

Such faculties were thought to result from unusual encounters with lightning, or with some other powerful electrical force. This was the explanation given for the strange powers of Henry Luegeman, a Bennetsville, Indiana, farmer struck by lightning while peeling tanbark:

Whenever a storm is about to approach Luegeman becomes highly charged. Flies which happen to alight upon him drop to the floor dead, while such things as small particles of iron or steel cling to his fingers.

During a recent storm he drew the blade of a case-knife between his forefinger and thumb, which so thoroughly magnetized it that heavy iron particles could easily be raised with it. When placed in a dark room thousands of tiny sparks are emitted from the man's body, and his eyes shine as brightly as if they were incandescent lights. Luegeman claims that he feels no inconvenience from his peculiar gift, but will go near no moving locomotive or heavy iron machinery in a storm for fear of being drawn against it and killed.[87]

Extraordinary encounters with electricity were magical partly for their capriciousness. Anyone might be selected by nature for a first hand experience as profound as anything trained men of science could devise and reserve to their own exclusive realm of experience. Such experiences also seemed superior to the plodding efforts of experts to capture predictable phenomena in their elaborate nets of explanation. Remarkable encounters with nature required no learned preparation, and bestowed a special mark of esteem, being born of a terror and dread no scientific investigation could approach. *Popular Science News* reported that Mrs. Charles Conover of Nanuet, New York, sitting in a piazza chair during a severe thunderstorm, had been shocked into unconsciousness for seven hours. The village doctor who examined her claimed that the shock had turned her heart upside down, in spite of which she was as well as ever.[88] A New Salem, Vermont, family, roused from slumber by a terrible crash of lightning, discovered that their cat had been electroplated with silver stripped from a Revolutionary sword hanging over the sofa where the cat had slept.[89]

Electrical miracles elevated humble people and gave them powers surpassing in popular estimation those of scientists and engineers, and experience that qualified them to dispute even received scientific authority. In 1889, H. M. Stevens of Boston advanced the opinion, much under debate because of a bill pending in New York to permit electrical executions, that electricity could not kill a man. Stevens told a *New York World* reporter that while inspecting an electric light plant, he had grabbed both the positive and negative brushes of a dynamo carrying fifteen hundred volts. Two doctors working on his "cold, stiff, and pulseless body" for three hours had been unable to revive him because he was, in his own words, "full of electricity and insulated." An attendant suggested putting him on the damp floor so that "the electricity could run out." This was done, the electricity "ran out," and "Mr. Stevens rapidly recovered and grew fat, but remained a sort of dynamo, and still is very sensitive when thunderstorms are approaching."[90]

People did, of course, die from electrical shock, and the gruesome manner of their going was presented as a bizarre and exceptional fate laced with magical and strange occurrences. Consider this account from *Popular Science Monthly* of 1873:

> Sometimes one struck by lightning is killed outright on the spot, the body remaining standing, or sitting; sometimes, on the contrary, it is thrown to a great distance. Sometimes the flash tears off and destroys the victim's dress, leaving the body untouched, and sometimes the reverse is the case. In some instances the destruction is frightful, the heart is torn apart and the bones crushed; in others the organs are observed entirely uninjured. In certain cases flaccidity of the limbs occurs, softening of the bones, collapse of the lungs, in others, contractions and rigidity are remarked. Sometimes the body of the person struck decomposes rapidly, but at times it resists decay.[91]

Camille Flammarion, a popular writer known for stories about visionary future worlds, was said to have "devoted considerable time to the study of the effects of lightning on men, animals, and other objects." Flammarion had collected a number of anecdotes about occasions on which natural electricity had acted with fantastic perversity. Such stories confirmed the mystery of nature and its presumed delight in thwarting scientific efforts to understand it, not to mention the superior virtue of those who refrained from every temptation to pierce the veil. Several of these stories, doubtless selected with an eye to their tingling entertainment value, were summarized in *Popular Science News*:

> During August of last year a young man was struck by lightning and carried a distance of 50 yards without being in the least conscious that anything unusual had happened to him until the lightning threw him against the wall of a house, where he received a slight wound in the knee. Two cows, which were 200 yards from him when the bolt fell, were killed.
>
> Three soldiers are said to have sought refuge under a lime tree during a storm, and all three were struck by lightning and killed as they stood side by side. Though dead, they maintained their erect position, but their bodies, when touched, crumbled away into dust.
>
> Two peasants were preparing to eat their breakfast, when suddenly the dishes were thrown to the ground, the bread, cheese, and fruit vanished from the table, and they were covered with straw.
>
> On another occasion a man walking through Nantes was suddenly enveloped by lightning, yet remained uninjured. When he reached home, however, and opened his purse, which had contained two pieces of sil-

ver and one of gold, he found that the gold piece had vanished and in its place was a silver piece. The lightning had, in fact, pierced through the leather of the purse and had covered the gold piece with a coating of silver taken from the other two coins.[92]

Soldier companions, slapstick straw-covered peasants, and foot travelers bearing modest worldly treasure in leather purses were colorful stock fairy-tale characters, in spite of the illusion of truthfulness created by the provision of realistic detail: the specification of the kind of tree beneath which the soldiers perished, the name of the city where lightning had transformed gold, the exact month and year and the specific distance that lightning had lifted the fortunate young man in order to spare him. Other fairy-tale elements were disguised in the vocabulary of scientistic explanation. Fate capriciously awarded life, death, fortune, and humiliation, but in terms acceptable to scientific realism. No trickster deity had robbed the traveler of his gold; it had only been silver-plated in exceptional circumstances. Magic remained, but was set down in a scientifically bounded world.

The public's fascination with freaks whose bodies exhibited mysterious electrical powers was not lost on carnival promoters and fast-buck entrepreneurs. Testimony alleged to come from the very mouth of a certain Johnny Norton, advertised as "Bonnell's Traveling Electric Boy" during the 1880s, recalled the successful and lucrative deception he had perpetrated. For exhibitions, Johnny stood in a stall separated by a counter from customers who passed one at a time in front of him along a narrow passageway. Concealed beneath the cocoa matting they walked across, and beneath the floor of Johnny's stall, was a continuous strip of zinc. One battery pole was located beneath the matting at the point where customers stood to greet the Electric Boy, the other beneath the Electric Boy's feet. Each customer who touched the Electric Boy completed a circuit, and both Johnny and the customer received a shock.

> It was surprising what intelligent people were duped by this trick. Why, I was kept shaking hands and being fingered from morning until night. Many's the two-dollar note I received from doctors and others for a couple of drops of my blood for analysis. In fact, my arms were covered by scars made by scientific dupes boring for my electric gore. One evening three or four young students came in to unmask me. One of them made a wager that he would electrify the audience the same way if he was in the box. I immediately invited him in, and he accepted the challenge. I then retired, but before doing so I pressed a hidden button that

cut off my wire. He, of course, failed, and ignominiously retreated after
being guyed unmercifully by those present. This proved me genuine to
the satisfaction of every one in that town, and I became famous. There
was lots of fun in the business, but I had to give it up, as the constant
strain caused by the battery was too much for me.[93]

Theatrical stunts in which electricity was made to spark alarmingly were common in traveling lectures from patent medicine scams
to scholarly presentations. Nikola Tesla, a Croatian émigré to the United
States, whose investigations of high-frequency, high-voltage currents
were instrumental in the development of alternating current systems,
and who dreamed all his life of achieving wireless transmission of
electric power, was premier among these traveling lecturers. Tesla was
well known for a visually spectacular trick of passing hundreds of
thousands of volts through his body "while flames flashed from his
limbs and fingertips" by means of a special induction coil named for
him.

"The mere description of Tesla's actual experiments reads like an
impossible fairy tale," declared the *New York Sun* in admiration. "Those
who have felt the 'shock' from a small battery of a single cell will be
astonished to learn that Tesla passes through his body a current 200,000
times as 'strong,' and yet lives to repeat the experiment."[94] "Who could
fail to be carried away by the ingenious enthusiasm of the lecturer, or
remain unimpressed in the face of the weird waving of glowing tubes
in the suitably darkened room, and the mysterious voice issuing from
the midst of an electrostatic field?" wrote the sober *Electrician*.[95] Tesla
had received the recognition of the scientific community for inventing
the rotary-field generator, which he exhibited before the American Institute of Electrical Engineers in 1888. Despite this and other reputable
scientific work, he was frequently criticized for making science sensate
rather than cerebral. He was often extravagant in predicting fabulous
inventions that he never delivered. Even the popular press treated him
by turns as a visionary and a fraud.

Enhancement of the human body was a typical spectacular function of electric light. Tesla used it to glamorize science at a level of
corporeality impressive to the most untrained lay mind. Women's bodies were also decorated with electric light, logically extending their
conventional cultural adornment and no less frequently their cultural
objectification. In 1884 the Electric Girl Lighting Company offered to
supply "illuminated girls" for indoor occasions. Young women hired
to perform as hostesses and serving girls while decked with filament

lamps were advertised to prospective customers as "girls of fifty-candle power each in quantities to suit householders."[96] The women were fed and clothed by the company, and customers were "permitted to select at the company's warehouse whatever style of girl may please their fancy." "Electric girls" embodied both personal servant, potent status symbol of a passing age, and electric light as ornamental object, a dazzling display of opulence that signified status in a new one. Divorced from the body, impersonal electricity would in time banish personal servants and make electric lighting essential and functional for all classes, instead of a badge of conspicuous consumption for one. In Kansas City, employees of the Missouri and Kansas Telephone Company organized a public entertainment in 1885 in Merchants' Exchange Hall that was graced by "an electric girl, placed on exhibition" along with a model switchboard and telephone exchange.[97] And William J. Hammer, Edison's talented electrical effects designer, dressed his small daughter May as the "Goddess of Electricity, with tiny Edison lamps in her hair," earrings, breastpin, and a wand tipped with a star containing a tiny lamp, at a lavish New Year's Eve party at his Newark home.[98]

Some entertainers adapted the electric spectacle to adorn their bodies in performance. Before an astonished audience in Sheldon, Iowa, in 1891, Miss Ethyl J. Davis of the Ladies Band Concert and Broom Brigade rendered a tableau of the Statue of Liberty, reported in detail by the local newspaper:

> Miss Davis stood on an elevated pedestal, her upraised right hand grasping a torch, capped with a cluster of lamps, which alone would have been sufficient to illuminate the entire room. A crown with a cluster of lamps, and covered with jewels, and her robes completely covered with incandescent lamps of various sizes and colors completed the costume. The lights in the hall were turned down and almost total darkness prevailed. As the contact was made bringing the electric lamps into circuit, the entire hall was illuminated with a flood of light which almost blinded the spectators, and Miss Davis, standing revealed in the glaring light, certainly presented a picture of unparalleled brilliance and beauty.[99]

"The Greatest Event in the history of Brookings, South Dakota," was the decription given by local newspapers to a Merchants' Carnival held in 1890 at the Brookings opera house, where various industrial enterprises were represented by appropriately costumed ladies. One of these was Mrs. E. E. Gaylord, wife of the manager and electrician of the Brookings Electric Light Company. To represent that up-and-coming branch of commerce, Mrs. Gaylord wore a crown of incandescent lamps and her dress was decorated with the same ornaments. The lamps were all properly connected, the wires terminating in the heels of the shoes. On the floor of the stage were two small copper plates connected to a small dynamo. When Mrs. Gaylord reached the plates the 21 lamps of her crown, banner and costume instantly flashed up and she stood clad in "nature's resplendent robes without whose vesting beauty all were wrapt in unessential gloom."[100]

Decorating the human body with electric light was something more than an arresting item in the catalogue of novel electrical applications, precisely because the universality of the physical body offered a secure reference point for cultural experiments with new and strange technologies. The body was a known medium upon which to inscribe religious, civic, scientific, class, or sexual messages with the new instruments of electricity; conversely, electricity glamorized the body, whose very familiarity robbed it of some of its impact as a message medium. The electric light as bodily embellishment, a human-scale variation on the electric light as public spectacle, was a communications phenomenon of the first order. It appeared even in jewelry. Luxury "flash" jewelry from Paris was described in 1888:

> Electric jewelry usually takes the form of pins, which are made in various designs. One such trifle copies a daisy, and has an electric spark flashing from the center, another is a model of a lantern in emerald glass, while a death's head in gold, with a ray gleaming from each eye, is a third. The wearing of electric jewelry necessitates the carrying about of an accumulator [battery] which represents a spirit flash, and is generally stowed in a waistcoat pocket. Brooches are made occasionally for ladies' wear, but as women have no available pockets, a difficulty arises with regard to the battery.[101]

Most such ostentation was the preserve of the wealthy rather than the poor, the avant-garde rather than the outcast. In Paris, "the reigning queen of fashion . . . the young Marquise de Belbeuf, a charming creature with a decided taste for the bizarre and eccentric in dress," wore a device that made her the center of attention on social occasions: "It is her fancy to enter a ball-room crowned with a wreath of autumn blossoms, not too bright in colors, and with a bouquet of similar flowers in her corsage. Presently she touches a secret spring and both wreath and bouquet are brilliant with electric light."[102]

If electricity for bodily embellishment was considered a fashion statement in Europe, the same impulse was regarded with condescending amusement when indulged by less admired cultures. Accounts of the fascination of nomadic Persian tribes with the wire that extended British telegraphy from England to India by way of Persia were offered to Anglo-American audiences to illustrate the futility of furnishing European technology to uncivilized peoples. Europeans saw only that native societies did not share their appreciation of the physical connection of the telegraph to European civilization, or grasp the notions of property and rationality that Europeans projected on electricity. Nor did they reflect on how their interpretations of encounters with indigenous groups always justified the imposition of colonial power on subject peoples who were deemed base, petty, and deserving of domination, or that the single most important lesson subject peoples were expected to learn from electricity was that European culture was superior to their own.

One account in this genre came from the pen of Thomas Stevens, a British telegraph operator stationed in Persia, where the nomads were fond of "embellishing their charms of person" with thick bracelets of telegraph wire. Copper and silver greatly appealed to them, but as unsophisticated primitives, they had no objection to bracelets of baser metal, especially if, in accord with Stevens's view of primitive psychology, these could be obtained "without pay." Nomads in the know "went strutting about the country wearing a wealth of telegraph wire bracelets that made the eyes of their less lucky tribes bulge with astonishment." Since obtaining the bracelets required only "climbing up the Ingliz poles and hacking off the wire," the new fad spread quickly among the Elianites, Susmanis, and Bactiaris. This unlimited boon to the nomadic economy was carried into increasingly remote regions, until the British finally appealed to the shah to protect the line. " 'Very good,' said the king of kings, blandly, 'it shall be stopped.' Orders were sent out to cut off the hands of people who were found wearing telegraph wire bracelets. This had the desired effect, and to-day the telegraph wires are as safe in Persia as in any country."[103]

Vanity, greed, and a narcissistic preoccupation with the circumscribed world of the savage body seemed to Britons to have supplanted the opportunity that they, as bearers of advanced civilization, had offered benighted peoples to lift their horizons to a level of British enlightenment made universal. If Britons regarded the despotism the shah exercised over the bodies of his subjects as barbaric, they also tolerated it with an amused nod to *noblesse oblige* which required at least a show of respect for the principle of authority, no matter how brutally administered, so long as European goals were accomplished.

Machine versus Man

Still another set of comparisons between electricity and the body coun-
terposed them to each other by asking whether men were fundamen-
tally different from electrical machines after all. Arguments on this
score were ultimately concerns about the status of the souls and bodies
of men in a mechanical world, and fantasies of a contest for supremacy
between automation and evolution. Perhaps evolution and automation
were separate but competitive races moving in the same direction; alternatively, perhaps men would be evolutionarily transformed in the
process of adapting to an automated environment. That men might
become different creatures because of changes in their lives made by
machines was in some sense the concern to which every discussion of
the future of electrical technology was addressed. Nineteenth-century
observers were especially interested in how men might change their
biological constitutions or their ways of waging war in response to
machine imperatives, and both experts and laymen wondered how man
measured up to electricity conceived as a supernatural or supercultural
form. Perhaps electrical machines were cultural artifacts superior to
man himself. Perhaps they were debased cultural forms. Or perhaps
they were a highly advanced form of nature, destined to drive man
from his fragile position in the cosmos, rather than to help him establish its security.

Automation and Evolution

The man-machine link found expression in metaphors and clichés that
likened electric circuitry to human social organization, and in analogies
between electricity and the circulatory and nervous systems. In a paper
before the British Association in Ipswich in 1895, A. R. Bennett compared commerce to the lifeblood of a nation. It was as if, he explained,
roads, railways, and waterways were "the arteries through which this
blood is conducted, while telegraphs and telephones may be compared
to the nerves which feel out and determine the course of that circulation, which is a condition of national prosperity."[104]
    Comparisons between men and machines were drawn in both di-
rections. Men were depicted as not quite perfect machines, and machines were endowed with nearly human features. In a souvenir issue
of *Western Electrician* marking the arrival of the twentieth century,

inventor Forée Bain described the human body as a machine whose efficiency would be dramatically improved by electricity:

> The old machine has a very low efficiency. The digestive organs and the organs of assimilation, which are the boiler plant, have an efficiency much lower than a steam boiler plant. The fact is that the old machine consumes too much vital energy in preparing the fuel (food) for assimilation. Nearly all of this work can be done by artificial means, and as this is a chemical process and as electricity promotes and hastens chemical activity, no doubt this agency will be utilized as an assistant to the natural forces of the body.[105]

The precision of human physiology, on the other hand, inspired fantasies of machines that closely resembled men and might be much like them. Visions of a world in which machines would replace men had existed long before the introduction of electricity. But now automata powered by electricity were built, as well as imagined. Many of them provoked a fascination like that aroused by human beings whose bodies were believed to possess special electrical powers. Nine years before he transmitted the world's first radio broadcast of speech and music, Reginald Fessenden described in the *Journal of the Franklin Institute* the construction of a doll with a simple short-term "memory" based on the principle of elastic hysteresis in metals.[106] When the doll's spring-controlled hand encountered an open flame, it would "remember" the flame for short intervals and withdraw the hand from subsequent presentations.

The electrical centerpiece of William J. Hammer's New Year's Eve party of 1885 was an electric automaton named Jupiter. To the strains of Professor Mephistopheles' Electric Orchestra, Jupiter presided over a supper of gourmet electrical dishes. Promptly at midnight,

> the thunderbolt pudding exploded, the sheol pudding blazed forth red and green fire, illuminating the room; the telegraph cake clicked messages, bells rang inside the pastry, incandescent lamps burned under the lemonade, while the coffee and toast made by electricity was rapidly "absorbed." The magnetic cake disappeared, the wizard pie vanished, Jupiter raised a glass labeled "Jupiter Lightning" to his lips and began to imbibe.
>
> The effect was astonishing. His eyes turned green, his nose assumed the color of a genuine toper, the electric diamonds in his shirt bosom blazed forth in all their glory and he shouted phonographically, "Happy New Year! Happy New Year!"[107]

A more attractive automaton was promised for the World's Fair. This would be a facsimile of Madame Patti, the famous opera diva, and would "embody her smiles, gestures and movements of the eyes." More wonderful still, a concealed phonograph loaded with cylinders would reproduce the prima donna's voice.[108] George Moulton of Bridgeport, Connecticut, was said to have invented a walking mechanical boy that "distorts its features to such an extent as to frighten the average woman, and makes as much noise as a railroad train."[109] The boy had a "colossal" head in proportion to his four-feet-eight-inch stature. Arrayed in Scotch tweed, he grasped a two-wheeled cart for transporting passengers. One arm was fitted with a start-stop switch, and the other with a steering rudder. Moulton claimed that his boy could push a man fifty-seven miles to New York in eleven hours. The first public exhibition of the boy was dramatically staged. Before a fascinated audience, two locomotive headlights were focused on a black curtain. The signal was given:

> At once a whirring noise issued from behind the black curtain. It was followed by the clanking of chains, the buzzing of machinery, and a din of banging and thumping that was deafening. The curtain parted and the mechanical boy walked in, pushing a cart. The head upon the figure was turning from one side to the other, and the facial contortions were ludicrous. The tin eyelids gave the eyes a singular appearance that startled some of the spectators. The boy began a circuit of the hall, lifting his feet and marching with a 19-inch tread.[110]

In the middle of the performance, a coil of wire sprang out from between the boy's legs and wrapped around one of them. Though his progress was halted, his head continued to turn and grimace alarmingly.

We are already familiar with the purpose of such performances. They represented still another effort by experts and showmen joined in various marriages of convenience to capitalize on the entertainment value of magic and the respect accorded it by the uninitiated. By using a magical framework of astonishing effects to present the latest scientific and technological achievements, experts hoped to win over their audiences and persuade them that experts made the best magic of all.

Electricity and Death

The sobering prospect that machines masquerading as obedient servants might absorb or conquer the creative striving that made human achievement possible and remarkable took a variety of forms. One was the fear of a level of prosperity at which the social division of labor might cease to impose clear class markers upon which an ideology of deserved individual fate, so important an organizer of social order, could be based. A more dreadful prospect was the ultimate dystopian possibility. In discussions of the newfangled war and weaponry that electricity was assumed to be on the point of introducing, there were occasional predictions that the entire race would meet destruction in an electric bloodbath that would turn back on its human perpetrators. "The historian who writes of the future war will turn the pages of Greek legends and smile sadly at Jove's smiting the lightning," prophesied an article titled "Visions of 1950" in *Technical World* magazine in 1911. Along the lines of the theme that electrical experts were inheritors of a glorious tradition, the article continued:

> The old War God hurling his thunderbolts will seem impotent beside man wielding the forces of nature for weapons. Magazines exploded without warning by darting, invisible, all-penetrating currents of electricity; devastating waves of electricity, or of some more powerful force, flashing over hundreds of miles consuming all that comes within their scourging blast. Guns, explosives, and projectiles will sink into the past, even as have the bow and arrow, giving place to howling elements clashing under man's direction.[111]

That all human life might be laid waste by mindless robotic instruments of destruction was at the other end of the imaginative scale from the optimistic hope that electricity would offer eternal secular life. Prophets of electric apocalypse differed as to whether this fate was a tragic consequence of the inevitable intraspecies struggle that was the flaw of the human condition, or whether in the last days men would find themselves victims of the machine as the unfathomable, hostile Other. Speculation about the role of electric power in future warfare appeared in a variety of forms, from popular fiction to official military reports. The bulk of military opinion reported in electrical journals was well represented in an 1890 memorandum from Lieutenant Bradley W. Fiske to American military authorities proposing that civilian support corps undertake all electrical work required for military operations. The next war, "when it does come," editorialized the London *Electrical Review* in behalf of Fiske's recommendations, "will be one in which science, and especially science, will play a most important part."[112]

Actual as opposed to fantasy developments in electrical warfare were mostly in the realm of communications rather than destructive weaponry. Searchlights and internal telephone and telegraph communications systems constituted most military applications of electricity in the late nineteenth century. These offered interested observers a taste of more extravagant possibilities. In 1884 the American navy fitted out the frigate *Trenton* with its own incandescent electric light plant and arc searchlight for sweeping the surface and depths of the ocean for enemy ships and torpedoes. By 1890 electricity was considered a strategic military element:

> The faithful servant of man now not only guards the vessel from the inventions of the enemy, but aims and fires the guns, illuminates the sights that the aim may be sure, discharges torpedoes, measures her speed, is the most successful motor for submarine boats, and renders possible a system of visible telegraphy by which communications may be flashed against the clouds and understood at a distance of sixty miles.[113]

Military systems of telephone and telegraph communication were updated with new wireless technology. In 1898 all the forts in New York harbor were equipped with dynamos for searchlights, one of which was the sixty-inch searchlight used atop the Manufacturer's Building at the Chicago World's Fair. Other Atlantic coast forts, including Philadelphia, Boston, and Savannah, also undertook plans for electric defense.[114] In 1905 *Telephony* reported that the U.S. government planned to buy thirteen hundred telephone sets for military use in the coming fiscal year, five hundred to be used in the field and the remainder in coastal defense forts.[115]

A favorite imaginative sport of popular magazine writers and readers was to spin out hypothetical military scenarios in which some electrically ingenious device secured a strategic battlefield advantage. A writer for the *New York Evening World* described armies furnished with high-potential dynamos and engines capable of delivering a volley of electric shocks "of infinite variety" to approaching enemies.[116] Inspired by this idea, a reader suggested:

> Would not a shell containing bottled electricity be a terrible weapon of war? Say an iron shell, steel pointed or all steel, lined with lead and containing electricity as a storage battery does, the shell to be so made that the impact will set free the electricity. Would not a shell carry enough to do great harm to an ironclad?[117]

The *pièce de résistance* of conjecture about military applications of electrical technology was the fantasy of anonymous combat with remote-controlled automatic weapons, the ultimate in warfare at bodily remove. Even before Marconi's demonstration of wireless telegraphy in late 1896, the work of scientists like William Crookes and Oliver Lodge in detecting electrical oscillations at a distance had attracted popular interest and inspired a host of prophecies about wireless rays of destruction. One of these appeared in *Lightning* in 1896, announcing a new invention that would work "revolutionary effects on the art of modern warfare," and featuring the useful imprecision of detail that offered the most fertile ground for imaginative development:

> The precise details of the invention are for the present unpublished, but this much may be said. . . . By means of a readily portable apparatus, mechanical energy developed by powerful steam engines . . . is converted into wave energy in the luminiferous ether that pervades all space. This in turn by ingenious arrangements can be propelled in a continuous stream in any given direction, and caused to concentrate, or focus, at any desired point even many miles distant. There the energy is liberated with an enormous evolution of heat and a volcanic disruptive force that immediately annihilates every person and everything within an area that can be enlarged to any desired extent by a slight motion of the directing mechanism.
>
> So far as is at present understood, nothing can arrest this terrible though impalpable beam of destruction.[118]

Marconi's success several months later added shape and detail to contemporary visions of an ultimate weapon. Items that embroidered narratives of destruction on otherwise recognizable accounts of the new Marconi method of wireless signaling were widely published:

> [Marconi] claims to have discovered the means of blowing up the most powerful ironclad or torpedo at a distance of some twenty miles without the help of explosives of any description, his only stipulation being that the object of attack has a powder-magazine, and is heavily coated with metal—the thicker the better.
>
> His plan is this: He sets two spear-shaped wires on some high elevation, such as a turret or masthead. These are connected with a powerful dynamo fixed on the ground. So soon as the ironclad is sighted the points of the wire are focussed towards it, and the dynamo is set in motion. The electric waves discharged from the wires sweep through the air until attracted by the ironclad, which acts as an electric accumulator. After a few minutes, varying in length with the strength of the dynamo, the charge on the ship is so great that sparks begin to fly from all quarters, and particularly inside the powder-magazine, with its iron walls. The result need not be told.[119]

Not even wireless telegraphy was the culminating technology for mantic visions of destruction, however, since the emotional power of such phantoms attracted whatever the frontier technical advance of the moment seemed to be. When the Swedish inventor Axel Orling was credited with devising a shore-controlled electric torpedo powered by light waves in 1899, many popular periodicals reported that torpedoes steered by electromagnetic waves had already been secretly invented by Edison, Marconi, or Tesla.[120] An 1892 issue of *Answers* featured an item from an imaginary news report of 1992, impressed on phonograph cylinders and "read out" to the public: "As a result of the awful slaughter by artificial lightning sustained by both France and Germany in the recent, and what will probably prove England's last war, a Congress of nations will meet in Berlin to-morrow to arrange for a settlement of any future dispute by national arbitration."[121]

The terrible prospect of weapons invulnerable to pain and targeting victims without compassion inspired heartfelt doubt that rational men would allow their use. Against weapons that answered only to mechanical impulse and not to human fear or courage, men would be helpless. Since no victory could compensate for total destruction, war must cease forever. The age of peace was a historical necessity. By extension, some imagined it to be a historical inevitability. Invoking the serviceable pushbutton, R. F. Gatling, father of the Gatling gun, expressed his faith in horror as a deterrent at an inventors' convention in Washington, D. C., in 1891: "If someone could invent a powerful electrical machine that would kill whole armies by the mere turning of a switch . . . there would not be a monarch in the world who would go to war. He could not force soldiers to enlist if every man was certain of meeting death."[122]

Utopian authors painted world after world in which men had achieved this insight. *Answers* portrayed a millennial London, its traditional habits of communication, travel, and association transformed by electricity, where war was banned because of the simple knowledge, related by the Genius playing host to a nineteenth-century time traveler, that the power possessed by his society would depopulate whole provinces in a day if channeled into death and destruction.[123]

Edward Bulwer-Lytton's *The Coming Race*, published in 1871, described a future organized around a mysterious natural force called Vril, which closely resembled electricity in its physical properties. Its social achievements were the hopes Bulwer-Lytton's contemporaries attached to electricity. Vril was versatile and powerful:

> It can destroy like the flash of lightning; yet, differently applied, it can replenish or invigorate life, heal, and preserve, and on it they chiefly rely for the cure of disease, or rather for enabling the physical organism to re-establish the due equilibrium of its natural powers.[124]

The destructive powers of Vril defined the limits of social coercion. As an instrument of war Vril rendered pointless all superiority in numbers, military discipline, and technical skill:

> The fire lodged in the hollow of a rod directed by the hand of a child could shatter the strongest fortress, or cleave its burning away from the van to the rear of an embattled host. If army met army, and both had command of this agency, it could be but to the annihilation of each. The age of war was therefore gone.[125]

In sum, the image that appeared in many crystal balls trained on the future of war was a vision of dramatic technological change. But the collective response to this vision was startlingly passive. The unacceptable prospect of electric war would simply compel men to lose all appetite for it. Men aware of the consequences of pressing the electric button would choose peace.

Nikola Tesla gave this construction an unusual twist in a June 1900 *Century* magazine article, in which he argued that earlier advances in weaponry had not frightened men out of their bellicosity because no serious effort had been made to detach human passion from warfare. Tesla proposed to accomplish this by using electric technology to reduce the number of individual soldiers. For wars between man and man, he would substitute contests between machine and machine. If the work of war could be automated, war could be dehumanized— a result its twentieth-century inheritors can appreciate the full irony of. The mechanization of battle and the elimination of bloodshed would make "interested, ambitious spectators" of nations, a condition in which peace could be sustained.[126] To be a satisfactory and convincing surrogate for the grand exercise of human passion, a robot of war would have to perform its work "as though it had intelligence, experience, reason, judgment, a mind!"[127]

Like many of his contemporaries, Tesla believed that technology would act on fundamental habits of thought and practice, that it would transform deeply held social attitudes and inaugurate a new stage of human history. Despite the variety of their predictions about the direction of this change, Tesla and other observers had hit upon two central features of the social format of twentieth-century electric media: the remote connection between actual events and the audience observing them and an elaborate culture of vicarious experience.

Fiction writers were fascinated by the theme of remote warfare. In a Christmas story by John Kendrick Bangs, war was a profitable spectator sport at Madison Square Garden. A story in the British *Daily*

Mail portrayed twentieth-century war between Britain and Germany as a game in which organized verbal assault had replaced the usual physical confrontation. The humorous *casus belli* was the slogan "Made in Brighton," which Germany had stamped on its own cups, plates, mugs, and other souvenir goods. Battle artillery consisted of brazen trumpets. To the small end of each was attached a wire connected to a phonographic magazine. When its handle was turned, "voice force" was hurled through the air in the appropriate direction. As the belligerent armies met on the field, "Down with the German sausage-eaters!", "The British Bulldog Mangles the Eagle!", and other cries of abuse filled the air. The German gunners were considered the fiercest soldiers, "the guttural German voices and the uncouth German words being so eminently suitable to this form of attack."[128] Equipped with "the very latest" in speaking trumpets, the English infantry had a technological advantage and managed to win the war. Each force poured in "volley after volley of abuse, satire, epigram, and apt quotation, and here the superiority of the English made itself apparent."

Closer to home, concerns about using electricity for base purposes fueled a controversy in the late 1880s about whether electricity should be the agent of society's severest penalty against individuals. The focus of the debate was a bill introduced in the New York State legislature to execute Albert Kemmler, a convicted murderer, by electricity. After a sensational deathwatch in the press, the world's first electrocution ended Kemmler's life at Auburn prison on August 6, 1890. Press accounts dwelled on the grislier details provided by eyewitnesses, including the stench of the badly burned corpse where contact between Kemmler's body and the electrodes had been imperfectly made, and the convulsions and horrible sounds that were said to have issued from the dying man, who required two separate charges of electric current to be dispatched.[129]

Expert journals reprinted a broad range of expert and popular opinion, from expressions of outrage and revulsion to warm assurances that electrocution was efficient and humane in expert hands. The *Medical Record* questioned the notion that "to harness the lightning and bore it through a human body is thought to be one of the advances of the nineteenth century."[130] "The age of burning at the stake is past," editorialized the *New York Press*, "the age of burning at the wire will pass also."[131] A related objection held that electrocution was an ignominious and degraded application of electrical science.[132] Electricity was "too beautiful to be subjugated to such work as that," declared the editor of the French engineering periodical *Électricité*.[133] In remarks to the Electric Club, C. F. Brackett, a Princeton professor, deplored the "degrading of an agent which has done so much and is accomplishing more for the advancement of civilization than almost any other discovery in the history of the world."[134]

Many engineers and electrical entrepreneurs feared that any public association between electricity and death would endanger the industry's future. To use electricity to end life, Dr. Otto Moses of Lowell, Massachusetts, explained to a *New York World* reporter, "will certainly impair its value greatly for domestic purposes. If we make it an instrument of death, women and others will oppose its introduction into the household."[135] Precisely that result had been the object of a publicity campaign by Thomas Edison and his associates to discredit the alternating current distribution system of their rival, George Westinghouse. Much of the testimony that alternating current was an efficient means of killing people, and therefore the best possible instrument for electrocution, was orchestrated from the Edison camp, to the vast chagrin of George Westinghouse.[136]

Other experts were convinced of the comparative humanity of electrical execution, though they believed that simple bungling had made Kemmler's death unnecessarily protracted. With that bizarre mixture of attention to public awe of scientific magic, especially its most dreadful manifestations, and concern for craft precision, the editors of *Scientific American* had proposed the use of electricity for capital punishment as early as 1876, citing the deterrent effect of the superstitious folk regard for "death by lightning." The instrument they suggested was a powerful Ruhmkorff coil and a battery sturdy enough to last from execution to execution, all equipment to be operated by a "competent electrician" to avoid error.[137] Dr. Louis Balch, an electrical expert, favored the appointment of a certified professional to attend all executions and have charge of all electrical apparatus. "Had it been entirely in charge of scientific men," he declared, the Kemmler execution would have been smooth and flawless.[138]

Perhaps the most general question any epistemology addresses is the nature of the relationship between man and the world, what kind of world it is in which man finds himself, what counts as eternal verity and what is variable contingency in that world, and what level of control is possible in it. These were issues that gave conceptual anchor to discussions of electricity in both technical and popular literature, although different modes of inquiry cast electricity as either an extension of the natural world or its technological master. Attitudes about nature and the body provided scientists and laymen with distinct but flexible epistemologies for classifying and interpreting electrical phenomena.

The enormous range of discussion about electricity, nature, and the body attempted to locate electricity, a force of unknown dimensions, by means of the most familiar of all human landmarks, the human body. At issue was the relationship of electricity and technology to a larger natural and moral order, and the status of the human body as a probe and a point of reference for making strange electrical phenomena familiar. Specific cases addressed in popular and expert literature included intelligent electrical automata and electrical "freaks," or persons believed to possess special powers by virtue of unusual encounters with electricity; mythic encounters between embodied natural forces and the genie of human inventiveness; the moral propriety of electrocution; electricity as a healer or as a transmitter of bodily maladies; perhaps the destroyer of mankind or, alternatively, the key to life itself. Much of this debate was a competition between knowledge derived from oral-gestural modes, in which nature was treated as an ally with whom mankind was in direct dialogue, and knowledge based on formal, print-based theories that regarded nature as an object of conquest.

4

Dazzling the Multitude
Original Media Spectacles

> . . . imagine the stars, undiminished in number, without losing any of
> their astronomical significance and divine immutability, marshalled in
> geometrical patterns, say, in a Latin cross, with the words *In hoc signo
> vinces* in a scroll around them. The beauty of the illumination would
> be perhaps increased, and its import, practical, religious, cosmic, would
> surely be a little plainer but where would be the sublimity of the spec-
> tacle?
>
> —Santayana, *The Sense of Beauty*

Anthropologists and literary theorists are fond of emphasizing the par-
ticularistic and dramatic dimensions of lived communication. The par-
ticularistic dimension of communication is constituted in whatever of
its aspects have the most individually intimate meaning for us. The
dramatic dimension is the shared emotional character of a communicated message, displayed and sometimes exaggerated for consumption
by a public. Its dramatic appeal and excitement depend partly on the
knowledge that others are also watching with interest. Such dimensions
have little in common with abstractions about information and efficiency that characterize contemporary discussion about new communications technologies, but may be closer to the real standards by which
we judge media and the social worlds they invade, survey, and create.
Media, of course, are devices that mediate experience by re-presenting
messages originally in a different mode. In the late nineteenth century,
experts convinced of the power of new technologies to repackage human experience and to multiply it for many presentations labored to
enhance the largest, most dramatically public of messages, and the
smallest, most intimately personal ones, by applying new media technologies to a range of modes from private conversation to public spec-

152

tacle, that special large-scale display event intended for performance before spectators.

In the late nineteenth century, intimate communication at a distance was achieved, or at least approximated, by the fledgling telephone. The telephone of this era was not a democratic medium. Spectacles, by contrast, were easily accessible and enthusiastically relished by their nineteenth-century audiences. Their drama was frequently embellished by illuminated effects that inspired popular fantasies about message systems of the future, perhaps with giant beams of electric light projecting words and images on the clouds. Mass distribution of electric messages in this fashion was indeed one pole of the range of imaginative possibilities dreamt by our ancestors for twentieth-century communication. Equally absorbing was the fantasy of effortless point-to-point communication without wires, where no physical obstacle divided the sympathy of minds desiring mutual communion.

It is something of a historical irony that while late-nineteenth-century fantasies of perfect intimate communication are reasonably familiar, spectacles in which electric light played a widely admired communicative role have largely escaped our attention. Perhaps the reason lies partly in the fact that the electric light spectacle, emerging from inherited modalities—candles, bonfires, and oil lamps—was thought to point triumphantly in a future direction that, in the end, it did not. We think of the glittering spectacles with which modern audiences are familiar as something newer, invented by ourselves and perfected in cinema and television. On the contrary, elaborate visual spectacles were public occasions long before the introduction of electricity, and their transformation at its hands was very gradual.[1] Besides enhancing the effective impact of familiar gatherings, electric lights provided occasions for novel ones, such as the outdoor electric light spectacle and the ball game after dark. A number of familiar spectacles were in time done in by the success of electric lighting, mostly by focusing attention on the newer and more private spectacles they made possible. Several popular applications of the electric light, the fastest-developing of all commercial electric technologies, demonstrated its potential for mass communication. Sometimes it was simply a large-scale ornament for traditional spectacles. Other times it was a novel spectacle of its own, or an iconic representation of texts or figures projected on a cloud or a wall for all to see.

Both the actually occurring large-audience spectacle of the late nineteenth century and the ardently felt desire for intimate contact at a distance reflected a yearning to realize what then seemed fantastic ambitions for the possibility of human connection, a yearning that has fallen out unremarked during the long trek to a present in which we take both achievements for granted. Historically speaking, it is a deceptive present, since our achievements are not precisely those the late nineteenth century had in mind. That we no longer remember the excitement of electric light spectacles testifies both to the fact that mass communication was implemented more directly in other forms and to the tendency of every age to read history backward from the present. We often see it as the process by which our ancestors looked for and gradually discovered us, rather than as a succession of distinct social visions, each with its own integrity and concerns. Assuming that the story could only have concluded with ourselves, we have banished from collective memory the variety of options a previous age saw spread before it in the pursuit of its fondest dreams. Of course, our amnesia simply complements theirs. In just the same way that we often see our own past as a less-developed version of ourselves, the late nineteenth century projected its past as a simpler version of itself, and its future as a fancier one.

Enhanced Conversations

The smallest unit of communication is a dyad—speaker and listener—and we turn first and briefly to a fascination with perfect spiritual intimacy, inherited from a dream of centuries, which seemed to be on the verge of materializing in the late nineteenth century. The object of that desire was in striking contrast to the elaborately staged and brazenly public spectacles of light that were the crown of late-nineteenth-century electrical grandeur. Recurring anecdotes about electrically enhanced private conversations were often organized around the telephone, but the most fantastic projections concerned the prospect of eliminating all barriers, even wires, to the distant loved one. This communications topos was worked in endless variations, traced to antiquity, elaborated in a hundred hypothetical futures. The power of new electrical technologies to effect effortless intimacy between friends was perhaps the commonest of all prophetic themes about communication. Not incidentally, these predictions often portrayed experts and other interested commentators as heirs to the authority and achievements of classical culture. As reconstructed by the *Dundee Advertiser* in 1897, Paracelsus had recommended the following experiment:

Two friends who wished to converse at a distance proceeded thus: A piece of skin was cut from the arm or breast of each, and these fragments were "transplanted," so that either party had a portion of the cuticle of the other engrafted on his person. When separated from each other, at a given hour one of them traced on the piece of alien skin with a metal point the letters of the words in his message, and his friend could read these letters on his own arm, no matter how far they were separated.[2]

John Baptista Porta, the sixteenth-century founder and president of the first society for the physical sciences in Italy, elaborated this Paracelsian hypothesis by describing how two friends separated from each other could communicate with compass needles magnetized by the same lodestone, and thus remain sympathetically attuned forever. If one needle pointed to a letter on a dial, the other would automatically point to the same letter on a similar dial, no matter how far away. Porta's proposal appeared in a treatise on magnetic phenomena in a twenty-volume work on natural magic. "To a friend, who is at a far distance from us, fast shut up in prison," he wrote, "we may relate our minds; which I do not doubt may be done by two mariners' compasses having the alphabet writ about them."[3]

By the late nineteenth century, anecdotal tradition had transformed one alchemist into another. Thomas Edison was usually credited with devising a machine to render communication between friends at opposite ends of the earth possible without wires or appliances, as in this description of an apocryphal invention attributed to him:

Your friend abroad carries a small machine of this new invention, in size and shape resembling an ordinary watch. You carry a similar one. When you wish to communicate with your friend, you take out the watch, the needle of which is in electric sympathy with his machine. The needle oscillates like that of a compass, and when you find the direction in which it points you turn in that direction and think hard. That is all. The claim is that concentrated thought will produce an electric current, and that the mechanism of the new invention is so delicate that it will respond to this current.[4]

Mulling over a visionary promise of distant visual communication in an exhibit at the Electrical Exposition in Paris in 1881, the *Electrician* speculated on effortless long-distance communication in that mode:

The telephotograph of Mr. Shelford Bidwell even gives us the hope of being able, sooner or later, to see by telegraph, and behold our distant friends through the wire darkly, in spite of the earth's curvature and the impenetrability of matter. With a telephone in one hand and a telephote in the other an absent lover will be able to whisper sweet nothings in the ear of his betrothed, and watch the bewitching expression on her face the while, though leagues of land and sea divide their sympathetic persons.[5]

In 1897 the popular science magazine *Invention* told its readers that Nikola Tesla had sent wireless telepathic messages through twenty miles of earth, "and he still maintains that it is possible to send messages in this way to the Antipodes."[6] The same classical theme appeared in fiction. A lighthearted fantasy of life in the year 2099 combined the usual traditional elements with a topical theme of automatic pushbuttons. Its narrator explained: " 'We have wireless telegraphy, you know. Every man carries his own battery in his waistcoat-pocket. We put our finger so'—ting went a bell—'upon the top button, and call up our friends in any part of the earth.' "[7]

The changing raw material of scientific discovery made it possible to mix reality and fantasy in equal proportions of the familiar and the novel, and to rework the classical topos into ever more technologically precise visions of intimacy. In 1892 William Crookes offered the prospect of "telegraphy without wires, posts, cables, or any of our present costly appliances," not as the "dream of a visionary philosopher," but an achievement within the reach of existing technical knowledge.

> Any two friends living within the radius of sensibility of their receiving instruments, having first decided on their special wavelength and attuned their respective instruments to mutual receptivity, could thus communicate as long and as often as they pleased by timing the impulses to produce long and short intervals in the ordinary Morse code.

By tuning their instruments to identical wavelengths, "correspondents" could preserve the privacy of their exchange. Since for years before and after its invention wireless telegraphy was deemed useful only for intimate intercourse, this lack of secrecy was considered its chief drawback, as even "the most inveterate would surely recoil from the task of passing in review all the millions of possible wave-lengths on the remote chance of ultimately hitting on the particular wave-length employed by his friends whose correspondence he wished to tap."[8]

The actual invention of wireless communication in 1896 inspired this presentation to a British Imperial Institute audience by W. E. Ayrton the next year:

> There is no doubt the day will come, maybe when you and I are forgotten, when copper wires, gutta-percha covering and iron sheathings will be relegated to the museum of antiquities. Then when a person wants to telegraph to a friend, he knows not where, he will call in an electromagnetic voice, which will be heard loud by him who has the electromagnetic ear, but will be silent to everyone else, he will call, "Where are you?" and the reply will come loud to the man with the electromagnetic ear, "I am at the bottom of the coal mine, or crossing the Andes, or in the middle of the Pacific."[9]

By 1889, when E. H. Hall, Jr., the innovative vice-president of American Bell's long-distance subsidiary, AT&T, was ready to discuss his company's goals for the telephone, his terms were already familiar to his audience:

> We all appreciate the advantage of speaking face to face above all other methods of communicating thought because ideas are then conveyed in three ways—by the words, by the tone, and by the expression and the manner of the speaker. Until Bell invented the telephone only the first method was available—first by letter and later by the rapid letter—the telegram. It is perhaps fair to say that we obtain our impressions from all three of those things in equal proportions. I would not set bounds to the possibilities of inventive genius. Some day we may see as well as hear our distant friends when we communicate with them by the telephone.[10]

In the twentieth century, the telephone and television are both pervasive media of effortless communication at a distance. Of these two, the telephone is our favorite intimate medium, wires notwithstanding. Television, at least as we know it now, is not communication on demand with a cherished other, but, on the contrary, a ringside seat at the grand but impersonal spectacles of the world stage. In modern television, the element of the spectacle recalls the electric light show, the most dramatic tradition of electrical experimentation in the late nineteenth century, itself an enhanced version of earlier public spectacles. It included both the lavish exposition whose express purpose was to celebrate electrical achievement and the public occasion convened for other purposes, but extended and made more exciting by the decorative drama of electric light. It is to these spectacles that we now turn.

Enhanced Spectacles

It is instructive to recall Marshall McLuhan's argument of some years back that every medium "shapes and controls the scale and form of human association and action" surrounding it by virtue of its special physical configuration and capacities. According to McLuhan, each medium also conveys a semiotic eloquence about the culture of which it is part, which is expressively louder than any particular message content. McLuhan's notion of an information medium was unconventionally broad. In one of his best-known and most provocative examples, he insisted that even the electric light is an information medium.[11] As it happens, this would have been a perfectly sensible claim in both Britain and the United States in the late nineteenth century, although understood somewhat differently than McLuhan intended. Whatever the salience of the electric light to current perceptions of the media environment, it was thoroughly familiar in this role to late-nineteenth-century observers.

Its legacy survives in gay electronic signs that light up urban streets each evening, in lighted scoreboards in gymnasiums and on playing fields, in traffic signals, and everywhere else that lights convey intelligible meaning to audiences. But this was not all. In the perceived novelty of its high-drama public role, the electric light also expressed the sense of unlimited potential that was a staple of nineteenth-century discourse about the future of electricity. For if electricity was the star of the nineteenth-century show, its most publicly visible and exciting agent was certainly the electric light. It was present in exhibitions, fairs, city streets, department stores, and recreation areas. It was physically and symbolically associated with whatever was already monumental and spectacular. It appeared in grand displays, processions, buildings, and performances. It borrowed from every established mode of dramatic cultural self-promotion.

Lighting for special occasions was indeed the most visible and aesthetically indulged side of the new electric revolution. Spectacles were the fun that experts unashamedly allowed themselves in public, fun that was also dramatic testimony to their civic contributions as electricians and scientists. Spectacles were an opportunity for popular audiences to display enthusiasm for electrical science and entrepreneurship, and for public officials and electrical experts to make common cause. What was new was the need for expert personnel to mount these spectacles, and the promise of profit they held out to private entrepreneurs. Less often acknowledged, although no less present, was the sense of theater deliberately cultivated by these professionally inspired and directed occasions, and the ideology of power they projected.

One story may illustrate. For Peace Jubilee festivities to celebrate the conclusion of the Spanish-American War, the city council of Chicago appropriated two thousand dollars to erect a grand municipal arch at La Salle and Washington Streets. It was discovered that aldermen were forbidden to earmark city funds for this purpose. In response, the Edison Electric Company, Western Electric Company, and General Electric Company joined forces to erect an arch of "striking and brilliant effect" modeled on the triumphal arches of the Roman Caesars and received by the city with great fanfare.[12] Such favors were not soon forgotten. What electrical innovators and entrepreneurs could do for those in need of memorable displays of prestige made them useful friends to those in power.

The appreciation of electric light spectacles required no textual training, and electric lights were a professional achievement for which electricians were happy to be praised in the most extravagant terms. Technical discussions in the usual journals conferred scientific and technical legitimacy on entertaining and ideologically powerful social drama. Unlike discussions of other domains of electrical expertise, these were rarely accompanied by stories that rationalized the position of experts vis-à-vis others who misunderstood them or coveted their position. The very existence of these spectacles proclaimed cooperation between officials and experts and legitimacy for electrical endeavors.

The community ideology for which electric light spectacles were a vehicle was already well established by earlier modalities. Compared to the village bonfire that predated the Middle Ages, however, the float-wick oil lamp of the eighteenth-century garden fête, and even the carbon arc lamp, electric filament lamps seemed more colorful, elaborate, and versatile than other lights. Contemporaries claimed that their decorative illumination was more dramatic. The most immediate message conveyed by electric lights was that the occasion of their appearance was as colorful and as worthy of notice as they were. The association of experts with this new form of an old message confirmed their importance in every way. The purpose of *these* illuminated messages was not to intimidate heathen hordes, but to remind attentive audiences of the existence and justification of vested power, and to impress on them its size and majesty. This was communicated in the quality of wonder excited by displays of electric light, in their lavish scale, and in their clear and direct association with municipal and even national authority.

"Probably the most elaborate and extensive electrical illumination of streets ever attempted" in Pittsburgh was mounted in 1898 for the conclave of the Knights Templar. According to the trade press, both civic and professional ideology were motives. One was expressed as local patriotism, the other as a desire to surpass all previous efforts to advance Pittsburgh's reputation as an electrical center. Civic and religious symbolism saturated the display. Its large, elaborate scale conveyed a presentational excess that signaled a professional eagerness to execute the ideological aims of sponsors. Its main feature was a series of twenty-nine metal skeleton arches, jointly constructed by three local companies—the Pittsburg Ornamental Iron Works of Allegheny, Carter Electrical Company of Pittsburg, and Allegheny Light Company. Assisted by the Shelby Electric Company, the municipal Electrical Bureau had erected its own triumphal arch of 1,685 incandescent lamps. A blaze of electrical signs decorated Pittsburgh streets with a riot of lighted icons. Among the largest, also by Shelby Electric, was a mammoth Christian cross on the courthouse tower. The 110-foot-high, 60-foot-wide cross was outlined by a double row of 688 red lamps above illuminated letters made up of 2,300 frosted white lamps. Additional figures included Masonic crosses, shields and crowns, "waving flags, a knight on horseback revolving on a vertical pivot, interior figures made of miniature lamps and buried in palms, circles of lamps under fountain sprays, etc."[13] Whether the occasion was sacred or secular, electrical companies stood ready to demonstrate their versatility.

The symbolic possibilities of electric light attracted the interest of a variety of social groups that were eager to experiment. When the citizens of France presented Bartholdi's Statue of Liberty to the American people, "only the electric light [was] thought of" to illumine it. *Electrical World* declared that all serious propositions made by electricians would receive serious consideration from the government, and detailed a number of proposals that had already been made. One electrician suggested vertical beams of light projected upward from the torch, visible as "a pillar of fire by night." Another wished to have lights placed "like jewels around the diadem," and another suggested placing them at the foot of the loggia to light the entire statue, "and thus the illumined face of Liberty will shine out upon the dark waste of waters and the incoming Atlantic voyagers."[14] The final plans called for five electric lamps of 30,000 candlepower thrown heavenward to illuminate the clouds at a hoped-for visibility of one hundred miles in every direction, four lights of 6,000 candlepower at the foot of the statue to illuminate it, and incandescent lamps on the diadem to give a jeweled effect.[15] In every proposal the lighted statue was much more than a useful coast beacon; it was an icon of cultural grandeur.

An icon, moreover, that could compete with those of other cultures. In 1881 a correspondent for the *Electrician* thrilled at the nightly illumination of the granite obelisk in the Place de la Concorde for the Paris Exhibition.

> Every night the great electric lamp on top of the Exhibition in the Champs Elysée [sic] was projected on the monolith, and caused it to stand forth in the darkness like a spectral finger pointing to the stars. The day is not far distant when the ruins of the Colosseum at Rome, or the great temple of Carnac, will be illuminated by the glowing carbons for the wonder and astonishment of tourists.[16]

Cultural grandeur was also the message intended in the Paris Exposition of 1900 by the illuminated tableau before the Palace of Electricity, the main electrical exhibition building. Above the palace, allegorical statuary represented the Genius of Electricity, from which rays of multicolored light shone forth. In front of it, an Electric Cascade poured streams of water into a basin containing another illuminated sculptural group "representing Humanity conducted by Progress toward the future, and overturning into the stream two figures of Furies, representing the Routine."[17]

But if on some occasions electric light projected public order and embodied it metaphorically, on others it strained some essential definition of the situation. In the south of England in 1896, a Jewish synagogue that had installed electric lights was criticized for violating the Sabbath. Electric lights aroused suspicion at the traditional celebration of St. Damasus in the Vatican in 1885. Amid two hundred chandeliers with two thousand tapers burning in the temple housing the saint's remains, a stream of electric light shone down from the cupola on the pontifical tiara and keys above the canopy over the body of the saint. This rich decoration was denounced as profane and accused of resembling an especially lively scene from *Excelsior,* a play being performed at the nearby Costanzi Theatre. Distrustful ecclesiastical authorities prohibited the use of electric, calcium, and magnesium lights for sacred celebration, but by 1892 the Vatican had been fitted with electricity by order of Pope Leo XIII.[18] In 1894 the Church of St. Francis Xavier in New York City celebrated Christmas with three thousand electric lamps scattered throughout the church. Lights decorated the altars, the aisles, the statuary, and the pulpit. Additional effects included a lighted vase of porcelain lilies atop a lighted alabaster column.[19]

But the most enduring contribution of the electric light spectacle to modern mass communication was to the syntax of popular mass entertainment spectacles and the reorganization of their traditional audiences. In the first instance, the electric light contributed to the vocabulary of effects that defined the public spectacle. In the second, the growing popularity of domestic illumination, partly stimulated by the demonstration effect of the spectacle, helped transfer certain features of those spectacles indoors to private audiences. The effect of electric light on the form and content of the new kinds of gatherings it illuminated makes the modern television special, for example, quite as much the fruit of the electric light as of wireless telegraphy or the kinetoscope, and the electric light as much a direct ancestor of broadcasting content as the telephone is of network programming distribution. The same imaginative impulse that mounts the striking visual effect and larger-than-life excitement of a television spectacle today seized on the electric light for much the same purpose in the late nineteenth century.

Public and Private Light

Early in its career, the electric light was not well known as an illuminator of private spaces, the most familiar to us now of all its functions. There were no power lines, no wired houses, no technical substructures to support private installation. That development was more gradual, and much remarked. *Electrical World* declared in 1885 that gas still held sway in domestic lighting, and the British *Electrician* pronounced the electric light "at this moment a luxury."[20] As a form of conspicuous consumption for the well-to-do, domestic installation was cause for comment. In 1886 a list of noteworthy domestic installations in Europe included the residence of the chief engineer of the British post office, William H. Preece ("probably as perfect in its detail and interesting as any plant of this character," an observer commented);[21] the castle of the Marquis of Bute (illumined by 400 glow lamps); the Château de Ferrières of Baron Alphonse de Rothschild; the Java palace of Prince Manko-Negoro; and C. P. Huntington's mansion at Astely Bank, England.[22] In England, the *Electrician* singled out Didlington Hall, Brandon (150 Swan lamps); Warter Priory, near Hull (180 lamps); and Avenue House, Finchley (110 lamps).[23] When *New York Tribune* editor Whitelaw Reid purchased the mansion of railroad magnate and financier (and stockholder in the Edison Electric Light Company) Henry Villard, Reid's installation of permanent electric light fixtures without any provision for gas, in case the electric light plant failed, was considered highly unusual.[24]

The Paris Opera House was the site of the first practical electric arc light in 1836. In 1858, arc light enabled eighteen hundred workers at the Cherbourg docks to labor by night as well as by day. Not until the International Exhibition of 1878 did the illumination of several Paris streets with the newly invented Jablochkoff candle suggest the urban possibilities of electric light, which was promptly installed on the Thames embankment in London.[25]

In 1880 Wabash, Indiana, boasted of being the first American town to adopt the electric light for general illumination.[26] In 1878 Wanamaker's department store in Philadelphia installed Brush arc lights, and the next year Macy's was "brilliantly illuminated with electric [arc] lights until 10 o'clock," for which it substituted softer, more dependable incandescent lamps in 1883.[27] The first hotel in the United States to be lighted by electricity, the Hotel Everett on Park Row, blazed forth in 1882 with 101 incandescents in its main dining room, reading room, lobby, and parlors.[28] By 1895 electric signs in New York stood out "boldly in letters of fire" and formed "a prominent feature of the city's illumination after dark," according to the *Evening Sun*.[29] Most such signs were in front of theatres. Some changed with each program; others permanently displayed the name of the establishment to which they belonged. The most commonly used letters were twenty-five inches high, five lamps for each vertical stroke, four lamps for each horizontal one.

By 1881 nearly all the principal railway stations in London and many outside were lighted by electricity.[30] A year later the Pearl Street Station of the Edison Electric Light Company, the model for all early electric lighting systems, was operating in New York City. According to Thomas Edison, twenty-one American towns and cities had central stations actively supplying electric light by 1885.[31] Outside metropolitan areas, where central station systems were most common, isolated plants ran lights in factories, mills, refineries, workshops, and other manufacturing establishments. In fact, more than twice as many lamps ran on power generated from isolated plants as from central station systems. In 1886 an estimated quarter of a million incandescent lamps and ninety-five thousand arc lamps were operating in the United States.[32] Four years later the number of incandescent lamps exceeded four million. What had been five thousand central stations and isolated plants in 1888 had swelled to ten thousand by 1895.[33] New York alone was said to have five thousand arc lights nightly illuminating public places. Out in the western United States, tower arc lighting was a preferred option, the municipal light of choice in Detroit, Peoria, Elgin, San Jose, Los Angeles, Pekin, and Decatur.

According to the *Electrician,* the growth of incandescent lamps in London alone during the 1890s looked like this:

1890	180,000 lamps connected
1893	700,000 lamps
1895	1,178,600 lamps
1899	6,869,000 lamps[34]

Still, the introduction of electric light had to overcome not only economic but social resistance, not to mention ignorance. In 1885, Yale students who were getting "more light than they relish[ed]" chopped down an electric light pole erected at the corner of the campus. Student threats to dismantle the re-erected pole occasioned special police protection.[35] By 1896, the president of the National Electric Light Association, C. H. Wilmerding, estimated that 2,500 electric light companies in the United States and about 200 municipal plants represented an invested capital of some $300 million, and that 7,500 isolated plants represented an expenditure of perhaps another $200 million. From these 10,000 installations, 50,000 to 75,000 incandescent lamps were manufactured. By 1890, the National Electric Light Association estimated, upward of $300 million had been invested in constructing, equipping, and operating central stations.[36] By 1898, Boston led the country in the number of incandescent lamps per 1,000 people: 1,232. In New York the number was 859, and in Chicago 730.[37]

The Vocabulary of Electric Light Effects

The introduction of the electric light was everywhere dramatic. Often only nature itself, grandest of all spectacles, could provide a standard for comparison. When a small Illinois town installed electric lights, outlying farmers observing the glow were convinced that the town was on fire, and raced in their wagons to help. The electric light not only enhanced spectacles, it *was* a spectacle, captured in favorite descriptive metaphors of flowers, spider webs, lakes, and rainbows. A correspondent's account of the brand-new system of tower lighting in Detroit described each tower as a graceful structure as open to the wind and light as a spider's web. Five Brush lamps at the summit of each lofty tower shine down at night from the outer darkness, looking like so many crowns of living light shining above the city. . . . The five-pointed crowns of brilliant light, mile after mile along the avenue, far above the houses, and on a level with the weathercocks of the churches, give one the odd impression that Detroit has a celestial lighting apparatus of its own.[38]

Besides the comparison to nature, two other ways of rendering the emotional impact of electric lights were popular. One compared electric light spectacles to supernatural phenomena, fairylands, ghostly pointing fingers, and otherworldly, dreamlike settings. Another invoked the world of classical culture, clean, spare, and geometrically pure. All comparisons made it clear that the electric lights they described did not belong to the prosaic order of things. They belonged to a natural and supernatural world that seemed nearly, but not quite, beyond man's creative power, and to a legendary world of cultural giantism.

These worlds were often mixed. An electrical journal reporting on the First Greater America Colonial Exposition at Omaha in 1899 described its forty-five thousand lights as outlining "a veritable fairy city." A tropical garden by day was said to bloom again by night in "cunning clusters" created from more than three thousand electric lamps molded as roses, lilies, tulips, and other flowers, and nestled in real foliage: "a novel and brilliant work of art." Whirling globes of fire, festoons of colored light, broad shafts of a searchlight "now flashing here and there about the grounds, lighting up a pathway to the heavens above" were other details. To the observer reporting it for the electrical press, the complete effect was a benign "fairy scene, fleeting and unreal as the shadows of a dream." Above the white city after dark "will hang a misty cloud of light, playing on the heavens and shaming into shadow the twinkling stars. Viewed from a distance it might be mistaken for the light from a terrible conflagration but that it shines so clear and has no pall or pillar of dark smoke to mar its radiance."[39]

Despite the standard vocabulary of most descriptive efforts, the best effects were always those that had "never before been created" and were guaranteed to "astonish the visitors" that observed them, phrases that were bestowed on the plans of William J. Hammer, "the electrical expert," for the Ohio Centennial Exhibition of 1888. The *Cincinnati Times-Star* published those plans with relish:

Flashing lights will gleam behind the dashing waters of the cascade, there will be a fountain suspended by a single wire, the walks will be

bordered with changing lights; sprays of electrical bulbs will gleam up
from great vases of flowers—there will be lights, lights everywhere
blinking and gleaming, and flashing, now brightness here and darkness
there; then a total eclipse for a second, then a flood of light. A gorgeous
electrical rainbow will appear and disappear over the miniature Niag-
ara.[40]

The following year the peripatetic Hammer worked on another
"novel" exhibition for the Cincinnati Exposition:

It is a mammoth Edison electrical globe, composed of 15,000 of the
regular electrical globes. This exhibit will be nearly 30 feet high. It will
glow as one colossal lamp of 100,000 candle-power. It stands on a base
of about 3,200 globes that will gleam pure white, except where the word
"Edison" will flash and glow in colored effects.[41]

Though each new effect was "superior" to all that had come before, there was a vocabulary of typical light figures and effects. It
included stars, colored fountains ("an English invention and already
rather old," noted an American observer in 1890, "but a fine thing"[42]),
towers of light, globes, shields, flags, eagles, crowns, rainbows, illuminated arches, emblematic pictures, decorated foliage, floats, pools
and building facades outlined in light, white cities, outsized electrical
bulbs—and all in color, generally red, white, and blue, but occasionally green and amber as well. In a public lecture on "Practical Electricity" at the local library in 1892, the president of the Rutland Electric Light Company in Rutland, Vermont, demonstrated "artistic
groupings" of electric lights to which he gave the names "Emblematic
Banner," "Search-Light," "Midnight Photography," "The Bridal Veil
Fountain," "Wizard Circle," "Candelabra," "Spiral," "Whirlwind,"
"Electra's Arch," and "Electric Cooking." (In the "Electria Dance,"
the second part of the program, small girls fitted with electric lamps
of different colors danced to and fro on a stage among fir branches
laced with red, white, and blue lamps.)[43]

Electric light figures, treatments, and themes were similar from
occasion to occasion. Towers of light that demonstrated the adaptabil-
ity of electricity to decorative and theatrical effects were among the
displays essential to any ambitious exhibition featuring electrical inventions. The first electric light tower was said to have been constructed at the Lenox Lyceum in New York. Each successive tower
strove to be bigger and fancier than its predecessors. A particularly
glamorous account described the tower of light at the Minneapolis Exposition of 1890. Designed by Luther Stieringer of the Edison General

Electric Company, this tower represented an Egyptian lotus flower, icon from an enduring and powerful tradition:

> The pillar is the stem with its purple ribs at every angle of the octagon; the swelling top is the flower, with its petals and sepals strongly defined, the slender festoons of parti-colored light are the stamens—the whole a wondrous lotus flower. Many thousand incandescent lamps are used in producing wonderful effects.[44]

One of the most popular of these effects patriotically redirected the attention of the audience from a powerful but technologically distant culture to an immediate and progressive one. Lights were thrown on in quick succession while "The Red, White, and Blue" played, bringing the audience to its feet with a cheer. "The manner of producing this effect was wholly unknown to the spectators," the account added, "who looked upon it as almost supernatural."[45] Dramatic theatrical effect was as important a criterion of successful presentation as the more straightforward execution of technical plans and parts.

Lighted fountains were another popular device. "It has been discovered that by means of a few ordinary electric lights, the most ordinary water jet in a basin is converted into a fairy spectacle, the like of which even the pages of the 'Arabian Nights' Entertainments' make no mention of," explained *Electrical World* in praise of the fountains at the 1884 Electrical Exhibition at Philadelphia and the 1884 Health Exhibition in London, the latter said to be the most lavish application of electric light ever yet seen in England.[46] The principal feature of the Philadelphia Exhibition was an illuminated fountain, a "grotto-like" stone structure twelve feet high. Water from a pipe in the center sprang roofward "through a double ring of incandescent lights, and then, striking a central ball, fell back over them in a shower." Twelve smaller jets were thrown toward the center from the margins of the basin, surrounded by flowers and tropical plants, while three jets rose above all. At night multicolored arc lights played across the water and gave it the appearance of "a fiery fountain glowing with prismatic hues."[47]

Illuminated flags were still another icon. In 1896 President Grover Cleveland opened a national *saengerfest* at Pittsburgh by pressing an electric button in the Executive Mansion in Washington. The closed circuit lighted an immense American flag of "hundreds of beautifully colored incandescent lights" in a hall before an audience of eight thousand.[48] In 1892 dedication ceremonies for the World's Columbian Exposition in Chicago included a procession down Michigan Avenue past a number of illuminated displays constructed for the occasion. Among

them was a "most unique" decoration of two flags over the main entrance of the Auditorium Hotel. One was the royal standard of Columbus's patrons, Ferdinand and Isabella; the other the flag of the United
States. Together, they were studded with nearly three hundred incandescent lamps. Rows of light chased rapidly across the flags from right
to left "so as to give an impression to an observer on the ground that
the stationary flags were in reality waving in the wind," an especially
sought-after effect.[49]

In 1898 the *American Electrician* published complete instructions
for the flag-waving effect, as it was worked out in a Fourth of July
display on the third-floor balcony of the Palace Hotel in San Francisco
that was "admired by hundreds." The hotel's electric power was supplied by San Francisco Gas & Electric Company, but for the display
the management had placed "a large and well equipped laboratory at
the disposal of its electrical staff," a residue from the days when many
hotels had adopted electric lights to impress their status-conscious
clientele, and built their own on-site electrical plants. Here Edgar Drib-
ble, chief electrician of the hotel, arranged many novel effects for hotel
patrons. During that month, for example, the Christian Endeavor convention headquartered at the hotel was complimented with an electric
monogram of its emblem composed of more than six hundred yellow
and purple lamps.[50]

Searchlights were another standard item in the vocabulary of electric spectacles. A very early searchlight, a Maxim arc light, was projected from a tower on the Grand Union Hotel in Sarasota Springs,
New York, toward Ballston Spa, seven and a half miles away, in 1879.
Several hundred people gathered to witness this experiment. Its success
was measured by a test frequently quoted in reports of searchlight experiments. The test was whether a new technology of communication
could facilitate an old one, specifically, whether handwriting or news-
print could be read from the searchlight's distant glow.[51] "From the
Smithsonian tower the grounds are so illuminated that you can read
ordinary print," an electrical journal reported in 1884 about another
searchlight. It added that a watch face could easily be read a mile from
the Washington Monument.[52] During a week of opening festivities for
the Columbian Fair, a great beam of light in Jackson Park was "sent
circling about the horizon, up and down, fast and slow, concentrated
and diffused," with a Schuckert projector, one of the world's largest.
It was seen in Evanston, twenty miles away. A searchlight with a beam
visible to Milwaukee was promised for the Fair itself.[53]

In 1892 "the most striking effect" in the illuminated decorations for the Grand Army of the Republic encampment in Washington, D.C., was created with searchlights from four 24-inch projectors atop the Loan & Trust Company Building. At night streams of light from these projectors cast the national colors on the clouds or illuminated the surround of the city.

> The most marvelous effects were obtained, however, when the whole four were brought to bear on the capitol. . . . The nights being without moonlight, the background was absolutely black, and when the lights were thrown on the capitol it appeared magnificent and awe inspiring beyond conception. The illuminated monument was none the less effective. The white stone shaft springing up straight into the blackness of night, seemed like a supernatural finger pointing heavenward.[54]

The expert press gave its fullest attention to electrically enhanced spectacles that celebrated electrical achievement. But electric lights glamorized a variety of other spectacles as well, from the inaugural ball of President-elect James A. Garfield in 1889 to a week-long bicycle show in Chicago in 1898, where three hundred exhibitors with a multitude of illuminated signs advertised bicycle manufacturers. At the inaugural ball, more than 150 lamps trailed across arches in the main hall, 80 lamps decorated the pagoda at its center, 12 electroliers of 64 lights apiece were suspended from the ceiling, 6 brilliant arc lamps shone in the upper portion of the hall, and 2 arc lights on each of the four entrances to the building and on poles around it created a crowd-pleasing display.[55]

As attentive arbiters of the novel and the standard, electrical experts were also critics of conventional and routine effects. In connection with the Peace Jubilee demonstrations in Chicago, the electrical press courteously acknowledged the lavishness of illumination for the festivities, but observed that "no great novelty in design was attempted." The general effect was typical of civic displays. It featured a "great volume of light on the principal downtown streets from strings of incandescent lamps in overhead festoons and in a great variety of combinations on the building fronts and projecting signs and decorations." Conventional or not, "enthusiastic" spectators braved miserable weather for the greater part of a week to see it.[56] An October 11 pageant in New York that celebrated the four hundredth anniversary of Columbus's discovery of America disappointed the editors of *Electrical World* when it failed to live up to its billing. The editors added their own assessment of the public's threshold for thrills:

The announcements made in advance of the Columbus pageant held in this city last week led every one to expect that the electric lighting displays, both in number and variety, would excel anything of the kind ever before attempted. As a matter of fact they were few in number, and, with some notable exceptions, of a very inferior order. The time has passed when even the "dear public" cares for school boy displays of combinations of electric lamps that alternately flash out their rays of numerous colors. A number of years ago it was something wonderful, from the point of view of the general public, to arrange a set or a number of sets of incandescent lamps in combination with switches so that an almost endless variety of displays would be obtained. But a display at the present time must possess something more meritorious.[57]

Illuminated effects for encampments of the Grand Army of the Republic were standard not only in relation to the entire vocabulary of electric light effects, but also from encampment to encampment. The Grand Army centerpiece was always a large badge of red, white, and blue lights. At the Detroit encampment in 1891, this badge was forty-eight feet high and sixteen feet wide, with eagle, flag, and cannon emblems inscribed within. The inscription "Hail, Victorious Army" appeared in seven-foot-high letters spelled out with six hundred lamps, while the G.A.R. monogram appeared in letters twice as large. An anchor represented the navy, and a horse's head symbolized the cavalry. Each took about one hundred lights, and all was visible for ten miles down the Detroit River.[58] Two thousand more lights mounted above City Hall created a glow for five miles in every direction.

At the 1892 encampment in Washington, D.C., another monster badge was the centerpiece, eighteen feet high this time, with a double border of incandescent lights within which eagle, cannon, and flag were outlined, but with the addition of the pendant star. Each figure was a different color. Along the parade route from Pennsylvania Station to the Treasury Building, sixty-three smaller corps badges were placed at intervals and composed of red, white, and blue lights. Above each badge, more lights spelled out the name of a notable battle in the history of each corps. Revolving stars changed shape and color, and lights were strung across bushes, trees, ferns, and flower beds on the White House lawn. The whole required special poles to support a wiring system that drew current from every available public and private dynamo in the city. When even more current was needed, additional streets were wired and socketed for projectors and revolving wheels brought straight from the factory. More than twenty-five thousand incandescent lamps and a hundred arc lamps were pressed into service.[59]

Besides forming figures and spelling out letters, electric lights constituted still another kind of message in the Grand Army decorations. Suspended at intervals along the parade route were oil portraits of leading Union generals, each surrounded by a double border of electric lamps.[60] Just as electric lights on the Statue of Liberty enhanced the monumental function of sculpture, they here amplified the honorific role of the distinguished oil portrait, along with the generals depicted in them. One private electrical display on a storefront for the opening of the World's Columbian Exposition included a portrait of Columbus framed in white lights. Above it was inscribed the lighted word "Welcome."[61] Lamp-bordered portraits of Queen Victoria were featured in the decorations for her diamond jubilee in London in 1897.[62] Light-studded oil paintings of battle scenes and portraits of the President and war heroes were also a major ornamental device of the Peace Jubilee municipal arch in Chicago.

The Chicago Columbian Exposition of 1893

Although later expositions claimed to have surpassed it, one of the most successful of all late-nineteenth-century spectacles was the Chicago Columbian Exposition of 1893, a standard for many later expositions in America and Europe. Twenty-one and a half million fairgoers paid admission, and free passes were distributed to six million more. Total receipts for the fair were said to be almost $14 million, compared to $8.3 million collected at the Paris Exposition in 1889, the last large fair to precede it.[63]

The electrical presence in Chicago was manifested in ninety thousand arc and incandescent lights, giant motors that operated the machinery of the exhibitions, a complete fair telephone system coupled to the Chicago exchange, emergency police and fire alarm telegraphs, and additional telegraph lines to the outside.[64] But it was probably the fair's dramatic exhibitions and visual displays that inspired the average visitor's feeling for the vigor and significance of electricity. The Edison Tower of Light, a many-splendored shrine to illumination standing eighty-two feet high at the very center of the Electricity Building, was one of the fair's most popular attractions. Rising from a polygonal base faced with mirrors, and surrounded by a circular row of columns supporting a squat-roofed pavilion, a circular shaft studded with five thousand incandescent bulbs sparkled and flashed in changing patterns of color. The shaft was topped by a gilded capital upon which rested an eight-foot replica of an incandescent light bulb constructed from thirty thousand prisms. Behind these prisms more rows of light bulbs glowed and flashed.[65] Outdoors, electrically powered fountains flanked either side of the Columbian statue at the head of the Main Basin. Strings of bulbs outlined all important buildings and waterways at night, and every evening the play of thirty-eight colored arc lights on the rising and falling jets of water created another of the fair's major illuminated attractions, at an estimated nightly cost to the Edison Company of a thousand dollars.[66]

Contemporary descriptions of the fair's nighttime appearance were worshipfully extravagant. Many employed a turn of phrase that sounds quaint to twentieth-century ears accustomed to an electrical world. A number of these descriptions invoked nostalgic worlds that electricity was decidedly not to recapture, a world of classical grandeur and a world of nature. One writer observed that each evening's electric fountain light shows produced illusions of "great flowers, sheaves of wheat, fences of gold, showers of rubies, pearls, and amethysts."[67] Another description of the nocturnal fair recalled classical Greece at its height and found it wanting:

> In a moonless night Athens hid her beauties, and for the time being might as well have been some dingy, hideous manufacturing town of to-day.
>
> Not so, however, with this modern Athens, for night is the time of her greatest splendor. . . . Thousands upon thousands of incandescent bulbs trace in delicate threads of light the outlines of the facades. The dome of the Administration Building becomes one blaze of glory. Blinding searchlights seem to make iridescent living things of cold, dead statues. The lagoon becomes a sea of dancing lights, edged by a ribbon of dazzling brightness. It is a fairy-land, an enchanted place.[68]

A third writer sensed an evangelical power in the onset of the evening spectacle:

> Under the cornices of the great buildings, and around the water's edge, ran the spark that in an instant doubly circled the Court [of Honor] with beads of fire. The gleaming lights outlined the porticoes and roofs in constellations, studded the lofty domes with fiery, falling drops, pinned the darkened sky to the fairy white city, and fastened the city's base to the black lagoon with nails of gold. And now, like great white suns in this firmament of yellow stars, the search lights pierced the gloom with polished lances, and made silvern paths as bright and straight as Jacob's ladder, sloping to the stars; or shooting their beams in level lines across the darkness, effulgent milky ways were formed; or again, turned upward to the zenith, the white stream flowed toward heaven until it seemed

the holy light from the rapt gaze of a saint, or Faith's white, pointing finger![69]

In our own world, where electric lights are prosaically utilitarian and unremarkably plentiful, such descriptions can be understood only as a response to something quite novel in the experience of delighted observers, the introduction on a new scale of the grand illusion, an effect that still defines success in modern mass media. The impact of the electric light spectacle is no longer freshly available in the same way to our imaginations, accustomed both to electric lights and to more fully elaborated illusions, as it was to those who saw it for the first time in the nineteenth century.

Outdoor Spectacles of Commerce, Indoor Spectacles of Status

Thus did most people first come to know the electric light through its public appearances, which were frequently commercial. *Electrical World* claimed that shopkeepers first thought of the electric light solely as an advertising device and only gradually realized its additional possibilities.[70] The *Electrical Review* recalled the impromptu spectacles of its first appearance in New York early in the 1880s:

> We can well remember when the first electric lights appeared in the New York shop windows and over the doors. It was looked upon as a mere experiment, the continuation of which would soon prove more trouble than it was worth, and the neighboring stores took no stock in it. Soon, however, it was discovered that it was attracting the attention of customers and the general public to such an extent that its users were compelled to enlarge their stock. Owing to the brilliancy of the light pedestrians could walk by stores of the same character lighted by gas without even seeing them, so attractive was the brilliant illumination further along. They clustered and fluttered about it as moths do about an oil lamp. That settled it; the neighboring stores must have it, and the inquiry and demand for the light spread apace until now, when, as soon as the electric light appears in one part of a locality in an American city, it spreads from store to store and from street to street.[71]

A number of commercial and recreational activities were organized expressly to exploit the possibilities of electric light. Nighttime trolley parties, for example, were conducted in electric cars especially chartered "to make a run of an evening for the mere pleasure of the motion" and decorated with electric lights to signify their special recreational function.[72] Chartered cars also took revelers to parks ablaze with electric lights, and to illuminated evening baseball games.[73] Hired trolley cars were differentiated into "ordinary" cars, and "a stock of luxuriously appointed cars, gaily decorated by day and brilliantly lighted by colored lamps at night, in which beauty and fashion can spend an afternoon or an evening speeding about city and suburbs." In a procession more than a mile long, more than twenty-five hundred celebrants traveled in fifty-four cars to a celebration sponsored by the Knights of Pythias in 1896 at Gardner's Park in Chicago. Three decorated open cars carried more than one hundred passengers to the suburbs of Chicago for dancing and refreshments in a trolley party sponsored by the Chicago Electrical Association in 1899. In Brooklyn the charge for chartering a plain "straw party car" was fifteen dollars for a day or evening ride. A "decorated and illuminated" car cost twenty dollars for the same period.[74]

A number of electric light spectacles in the 1880s found a setting in New York Harbor. From the shore of Staten Island, New York at night looked like "fairyland": "a thousand electric lights dancing from out a sea of inky gloom, with here and there a cross, and there a crown, near which fireflies of huge dimensions start here and there with phosphor fires aglow; the streets ashimmer with silver, with calcined towers lumined against the unfathomable gloom beyond."[75] On Staten Island, a colored electric light fountain in the St. George pleasure grounds was put through its paces and drew applause from opening-night spectators:

> At one moment it was crystal, the next roseate, then successively green, blue, purple, gold, and from time to time the tints would blend, harmonize, and contrast with new charms at every change. . . . Far out in the bay it could be seen, looking like a gigantic opal, illuminated by its internal fires.[76]

Viewed from Staten Island, the splendid spectacle of Brooklyn Bridge lighted by eighty-two arc lamps was said to have excellent commercial potential: "It is so beautiful, in a scenic sense, that one of the enterprising ferry companies contemplates having nightly excursions during the summer season, which it is intended to advertise as the 'Theater of New York Harbor by Electric Light; price of admission, 10 cents.'"[77]

Another site of special lighted effects was Madison Square Garden. The more than thirty-six hundred lights installed there in 1890 had increased to more than five thousand in a year. Lamps were arranged in pleasing patterns around the building and its turrets and porches, and a powerful searchlight mounted on the highest part of the roof was projected on prominent buildings from Harlem to the Battery.[78]

By 1911 a number of cities were using illuminated booster signs for civic advertising. Easton, Pennsylvania, claimed to have erected the first one in 1907. Others were soon found in Phillipsburg, New Jersey; Montgomery, Alabama; and Denver, Colorado. A single sign was shared by the adjoining cities of Bristol, Tennessee, and Bristol, Virginia. The sign at Easton, a rail crossing point, could be seen flashing its legend, "Easton—City of Resources," every evening from dusk to midnight from eight different railroads. In Denver, a great arch of light facing Union Station and blinking the inscription "Welcome" brought a whole class of similar displays to one observer's mind:

> As one comes from the main exit of the station the scene is reminiscent of some of the great world's fairs of a few years ago. That towering arch, outlined with countless electric bulbs, forms a frame for a vista of illumination which is restrained from being a riot only by the orderly beauty of the street lamps which stretch in two rows as a barrier to the countless signs which cover the buildings on either side.[79]

In sum, the electric light was a public spectacle before it was anything else, certainly before it was a common furnishing in private residences. Prior to that useful but less glamorous destiny, it entered private places as an object of conspicuous consumption for a variety of communicative and illuminative uses. Indoors, the incandescent lamp was safer, cheaper, and more versatile than arc or gas lighting. Its aesthetic effect was softer and more radiant. It had neither the disagreeable fumes of low-candlepower gas lighting or the intense glare and uncertain safety of arc lighting.* In recognition of these virtues, some outdoor electrical events gradually began to move indoors to smaller settings. At first many of these were events for the socially prominent. On such occasions, electric lights imperceptibly transformed large outdoor community gatherings into smaller private ones, as small, in the end, as a single family.

Some of the earliest transitions from outdoor to indoor spectacles

---

*In the words of its best-known inventor, "The incandescent light is the most perfect form of artificial illumination yet produced. It gives a light absolutely steady, without any deleterious properties. It does not burn the oxygen gas out of the atmosphere, nor give it carbonic acid gas, and it is very pleasing to the eye. The consumer buys his electricity measured on a meter, which registers consumption with far more precision than can be attained in any gas registering meter."[80]

provided fresh excitement for another traditional spectacle, the stage drama. Noteworthy theatres that adopted electric light in the mid-1880s included Prince's Theatre, the Savoy, and the Criterion in London; Prince's Theatre in Manchester; the Prince of Wales and Royal Theatres, Birmingham; two theatres in Glasgow; the Court and Palace Theatres, Munich; the Court Theatre and Town Theatre at Brunn; and others in Berlin, Vienna, Stuttgart, and other Continental cities. In 1884 *La Lumière Électrique* reported on theatres lighted exclusively by electricity in Milan, London, Havana, Boston, Brussels, Munich, Stuttgart, Manchester, and Budapest.[81] That year, the Paris Opera House was fitted with four thousand incandescent lamps, then the largest number in any European theatre.[82] In the United States, thirty theatres were lighted by electricity entirely or in part by 1886. They included McVicker's Theatre in Chicago, the Chicago Opera House, the Milwaukee Opera House, and in New York the Metropolitan Opera House, Third Avenue Theatre, Lyceum Theatre, and Miner's Theatre.[83]

Lights arranged in extravagant chandeliers were a focus of interest in some theatres; in others, they figured in new kinds of scenic effects.[84] The *Pall Mall Gazette* reported in 1884, "Never before has such an effect been seen as a stage darkened for a moment only to be instantaneously illuminated with hundreds of ballet girls in armor, and every point in their stage harness picked out with stars of electric light."[85] The inventor of this trick had mistakenly been arrested as a Parisian Fenian upon arriving at Charing Cross station with his costumes and machinery. Two years earlier, chorus ladies at the Savoy Theatre represented fairies in a new opera and wore Swan incandescent lamps in their hair. Silk the same color as the ladies' dresses covered the conducting wires in order to render them less conspicuous.[86] The staging of the ballet *Monte Cristo* at the Empire Palace Theatre in London in 1896 attracted attention for its illuminated waterfall and a cave scene with six hundred lamps arranged to represent sparkling gems. "But the main novelty is in the production of electrical effects on the dresses and properties of 24 coryphees and 10 men dancers, each of whom carries an average of 100 miniature improved electric lamps specially shaped to suit the designs of the dresses." A stage cloth on the floor conveyed low-voltage current to the dancers' shoes and the wands they carried.

> In the "Apotheosis" of the first act the combination of miniature lamps (over 7,000 in number) is made to represent a diamond throne, and jewelry in the form of crescents, tiaras, pearl festoons, lovers' bows, stars, crowns, etc.; the lamps in this case being constructed on the model

of cut diamonds and round or oval pearls, and finally, the whole number of 34 dancers with illuminated dresses are grouped in the background after having performed their dances unimpeded.[87]

As fancy lights were laid on for sumptuous balls, receptions, and banquets, and entertainers appropriated the electric light as an indoor performance prop, effects that impressed outdoor publics were made to communicate the special status of the sponsoring parties inside. Illuminated indoor entertainments were often simply smaller-scale spectacles staged by experts, performers, and the affluent. One such instance was an electrically lighted Christmas tree exhibited at the Foreign Fair at Boston in 1884. Luther Stieringer of the Edison Company had designed the "strangely beautiful" spectacle of three hundred lights arranged in twenty-four circuits, which could be turned on in a variety of combinations.[88]

In 1885 an electrified Christmas tree inaugurated a personal tradition of electrical holiday hospitality by Edison Company president E. H. Johnson at his Greenwich, Connecticut, country home, which was fitted with every conceivable "comfort, splendor, and luxury" in electrical devices. Electric power kept the tree revolving and turning groups of its 120 lamps on and off. Other lights concealed under pine shavings created the illusion of a grate heaped with glowing coals.[89] Another Johnson presentation was a Fourth of July garden party with incandescent displays on piazzas in front of the house, an electric pinwheel of red, white, and blue lights mounted on a flagpole, and electrically ignited fireworks. Popular music played from a phonograph, a gift from Thomas Edison.[90]

Even absent its holiday finery, the Johnson house was a permanent spectacle of electric light. A large three-story colonial mansion set on the highest coastal ground between Maine and the Everglades, it was described as "a lantern in effect—a beacon in the countryside." Among its features:

> The two great drives that wind through the greenery of the park are fringed with electric lights, a wondrous cluster of carbon loops hangs in a chandelier under the arched entrance to the mansion where the drives terminate, the great verandas around two stories of the mansion glitter with these lights, and the tower surmounting all blazes like a lamp above the trees. All through the house are faucets for throwing on the brilliant lights, sometimes in the oddest shapes. . . . An electric fountain blazes on the lawn, and scores of sunken lights, hidden in the grass, illuminate the tennis ground for those who wish to play at night.[91]

Because most private lighting was temporary and difficult to install, a number of companies, such as the New York Isolated Accumulator Company, sprang up to meet the demand. In the United States, the first use of storage batteries to furnish "temporary lights for special occasions" was in 1888, at grand balls held at the Fifth Avenue mansions of Ogden Mills and Cornelius Vanderbilt.[92] Giant batteries were charged at the factory and placed in the cellars. At the Vanderbilt ball, lights burned from 9:30 P.M. to 3:30 the following morning, without a single interruption.

Lights could frame an indoor spectacle such as a grand ball, or might be its focus, as in a novelty act created by Chicago entertainer George W. Patterson with multicolored lighted Indian clubs and "electrical spectacular effects." With the help of a portable generator, Patterson performed his act for many years in halls and churches. By swinging clubs in a dark room, he created the illusion of circles and other designs of solid light. Describing this act in 1899, *Western Electrician* detailed one of its striking features, the so-called electrical storm,

> beginning with distant heat lightning, gradually increasing to the fiercest of chain or "zigzag" lightning, with corresponding graduation of thunder, the latter being produced in the usual manner by a "thundersheet" of iron. . . . The effect is very startling, especially as it is accompanied by the fiercest thunder, the sound of dashing rain and by Mr. Patterson's voice laughing and singing "The Lightning King" through a megaphone. The "Lightning King" is followed by the latter part of "Anchored," in which a perfect double rainbow gradually appears and is dissolved by a water rheostat, by sending the rays of a single-loop-filament incandescent lamp through a prism. The colors come out beautifully.[93]

An intellectual version of the outdoor spectacle was the indoor electrical lecture with a variety of marvelous effects offered in the service of education. A lecture sponsored by the Edison Company in Boston featured a number of fascinating small-scale light effects, including a live goldfish with a lamp in its stomach, a model of the statue of Liberty Enlightening the World, advertising signs, red and blue dishes of water illuminated by incandescent lamps, a demonstration of military signaling from model balloons, and the large sign that accompanied all Edison Company productions, flashing out rapidly in succession the letters "E" in white, "D" in blue, "I" in orange, "S" in green, "O" in red, and "N" in white.[94]

As a member of the class of remarkable electrical inventions rapidly bringing about the future and as a bearer of messages to large audiences, the electric light seemed to many late-nineteenth-century observers to herald the very tidings of the twentieth. In spite of that fact, neither the electric filament lamp nor its closest competitor, the carbon arc, was truly a novel message medium. Electric lights extended earlier uses of bonfires, candles, and oil lamps to punctuate important nighttime social gatherings with spectacular effects, and earlier uses of signal lights to transmit news, particularly military intelligence. Electrical journals offered accounts of both military experiments with electric light semaphores and various improvisations. In 1898 in Wadena, Minnesota, the manager of the local electric light plant placed red and blue electric lights on the town's water tank to broadcast the battle-by-battle progress of the Spanish-American War.[95] During the 1890s lights over the dome of the Capitol in Washington, D.C., alerted observers that Congress was in night session. A red light denoted that the House was in session, and a blue light the Senate. When both houses met, a white light was added for full patriotic effect.[96]

Texts of Light: Advertising and Politics on the Clouds

We have noted that the impulse to dazzle audiences with electric light effects was not limited to entertainments in which the very presence of electric lights was the message of the spectacle, but was also expressed in inscriptions and simple figural representations constructed from lights. Antecedents for texts of light go back at least to 1814, to the construction in St. James's Park, London, of an illuminated sign in the design of a star and the words "the Peace of Europe," assembled from more than thirteen hundred spout-wick oil lamps. These were attached to iron frames and were intended (prematurely, as it turned out) to celebrate the end of the Napoleonic Wars.[97] Electric filament lamps made such achievements simpler, and inspired more ambitious ones.

At the wedding of Lydia Miller and David Rosenbaum in Baltimore, a variety of electrical effects decorated the reception. The bridal couple received their guests beneath an arch festooned with red, white, and blue electric lights, and lamps outlining the figure of a heart, the initials of bride and groom, and the year, 1892. When the guests reassembled in the supper room, a sudden outburst of bells and music was accompanied by a blaze of light. "At the completion of the first course the words 'Good luck' appeared over the heads of the newly-married couple and an electric hairpin, a gift to the bride, became incandescent and surrounded her head with a halo of light. Wine bottles were suddenly transformed into glowing candelabra."[98]

In 1899 a wedding spectacle in Atlanta featured illuminated textual decorations, and transformed one message into another before the audience's very eyes as lights flashed on and off. The groom, an electrical contractor, had set the wedding scene with two hundred incandescent lights draped from one side of the sanctuary to the other. Directly above the altar hung a wedding bell fashioned of foliage and one hundred colored lights. An arc light suspended from the interior of the bell represented the clapper. Further details were reported:

> To the right of the bell a letter N, the initial letter of the name of the bride (Miss Daisy Nimmo), formed of white incandescents, set in pink flowers, was supported on invisible wires. A letter L [for F. H. Lansdale, the groom] was on the left. . . . As the bride with her brother entered the church by one aisle and the groom with his best man entered by the other the letters N and L flashed into view and sparkled with great splendor. A murmur of admiration arose from the auditorium at the superb effect.[99]

Advertising and political spectacles that thrived on large audiences lent themselves to illuminated messages constructed on the large physical scale suited to group display. A common advertising device was the "sky sign" that spelled out a promotional slogan or the name of a firm, or projected an image against the blank wall of a building. At the instigation of Long Island Railroad president Augustus Corbin, anxious to increase his road's volume of passenger traffic, an especially ambitious sign was erected in 1892 on the side of the nine-story Flatiron Building at the convergence of Broadway and Fifth Avenue with Twenty-third Street in New York City. Fully lit, it consisted of fifteen hundred white, red, blue, and green incandescent lights arranged in seven sentences of letters three to six feet high. From dusk until eleven o'clock every evening, each of these sentences lit up in succession and in a different color, and brought "to the attention of a sweltering public the fact that Coney Island . . . is swept by ocean breezes," reported *Western Electrician*.[100] So long as the changes were "being rung," it observed,

> The public is attracted and stands watching the sign, but as soon as the whole seven sentences are lighted and allowed to remain so, the people move on their way and the crowd disperse. This illuminated sign is not

only a commercial success, but when all the lamps are lighted is really a magnificent sight. Its splendor is visible from away up town.

An early prototype of such messages was a signboard of gilded zinc with lettering stenciled out in punched holes, assembled by J. L. Blackmer of Boston and described in 1878 by *Scientific American*. A bright tin disk or glass lens behind each hole reflected light from the front. Each disk was vibrated by the moving armature of an electromagnet as a circuit to a battery behind the apparatus was alternately broken and completed. The gaudy effect offended the editors of *Scientific American*, who took the occasion to complain of the ubiquity of visual advertising in general. The effect on other observers, however, was exactly the same one that electric lights would soon inspire:

> We do not believe that any one can come within a block of that sign without being morally dragged into looking at it. A crowd, as we write, is standing open-mouthed staring at it. As an individual sign it is an astonishing success, and everybody who sees it will depart with the words "Homes in Florida" persistently flickering on his retina and shaking through his nervous system, dismally suggestive of the tremulous malady incident to Florida swamps. But then, supposing this sign came into general use; suppose both sides of Broadway united in one grand twinkle and flicker—the idea is too horrible.[101]

Even those who seemed to be sophisticated about electric light effects declared themselves hypnotized by serial message light signs that today strike us chiefly for the dullness of their content. A visitor to an exhibit of the Safety Insulated Wire & Cable Company at the National Electric Light Association Convention of 1896 related how his attention was drawn when "just in front of me there flashed out a sentence in red and black letters on a white ground. . . . A moment before the picture was a blank, but now I was challenged to read it; had to stop, and before I had finished there was another little picture on the side wall which told a story and disappeared." As he stood watching the most effective piece of advertising in the entire show, an acquaintance "who had charge of an electric light plant in a suburb of the city" happened by, and was similarly transfixed as a facsimile of the heading on a Western Union telegraph blank appeared in lights. The words "Mr. and Mrs. Delegate" were spelled out, followed by more words, one after another: "You—are—cordially—invited—to—visit—our—factory. Plant—is—extensive—and—process—interesting. Safety Insulated Wire & Cable Co., 225 to 239 West Twenty

eighth street, New York City." The advertising man who recorded this
scene estimated that the impression it made was "10 times more for-
cible [than] if they had all been shown together, perfectly constant and
stationary."[102]

   To attract crowds patrolling the streets of large cities, advertisers
often used electric lights and magic lanterns in combination. On "Magic
Lantern avenues" in Paris, commercial messages were projected on
shopwindows high above the street. In London "diverting and artistic"
displays were found in the Strand and Leicester Square as early as
1890.[103] The magic lantern also projected "living photograph" advertisements on the hoardings, on pavements, and even on Nelson's Column until the Office of Works prohibited this "desecration" in 1895.[104]
In Edinburgh, an electric signboard in front of the Empire Palace Theatre flashed out in 130 colored lights the words "Empire" and "Palace"
alternately, so that one took the place of the other.[105] In Sacramento,
California, "Mon. Leak" was said to be building a railroad advertising
car to trail across the countryside carrying sample goods inside and
advertising signs outside. The whole outfit would be brilliantly illuminated. "The car will make a mighty show bowling along after dark
through a farming region," observed the *Electrical Review*. "Wonder
what the farmers will think it is?"[106]

   State ceremony was another electric light occasion. In 1897 an
English electrical journal described some of the electrical tributes planned
in connection with the celebration of Queen Victoria's diamond jubilee, including a decoration, sponsored by one of the great cooperative
stores, of an eight-rayed star with five 8-candlepower lights on each
of its arms. "When the wheel is revolved the whole 40 lights merge
into a quivering sheet of flame, the outer edges of which acquire the
prismatic hues of the rainbow."[107] (On another occasion, this one a
royal birthday, the queen had summoned an electrician whom she commanded to outline all Windsor Castle with incandescent lights for payment of fifty pounds sterling. Having estimated that such a display
would cost fourteen times as much, he tactfully suggested a crown of
arc lamps around the top, "and by using scrap boiler and some old
engines, managed to keep the expenses sufficiently low to meet the
needs of even the Queen of the richest country in the world.")[108] The
London & Northwestern Railway planned to decorate its offices for
the jubilee with three thousand lights in a design that included a jeweled crown over the central dome, the letters "V.R." (Victoria Re-
gina), and a motto—"Longest, Noblest, Wisest Reign"—that repeated

the initials of the firm's name. Still another of the devices planned was said to have been employed "with great effect at Paris during the fêtes given in honor of the czar of Russia." It consisted of incandescents contained in varicolored celluloid balloons: "When these spheres, gently glowing with blue, red or white light, are hung in garlands, built up in pyramidal form, or are made to pick out the various portions of a Union Jack, the effect is decidedly novel."[109] Additional decorations included lamps screened by cut-crystal letters, floral designs with glass shades (to represent colored glowing blossoms surrounded by metal foliage), coats of arms, and the imperial crown. The same year, Berlin celebrated Emperor Wilhelm's birthday with a "grand illumination" by stringing banks of multicolored lights, arranged in the initials of the emperor and empress and the significant dates of their reign, across public buildings and private houses.[110] Many of the decorations derived from an international vocabulary of electrical effects that moved from country to country, occasion to occasion, cliché to cliché.

One-time political events also appropriated the drama of the electric light. Popular feeling kindled by Admiral Dewey's return from the Philippines in 1899 found expression in lighted displays of public support across the United States. In New York, the scene of the admiral's triumphant homecoming, an enormous "Welcome Dewey" sign in lighted letters 36 feet high was stretched 370 feet across the Brooklyn Bridge. The letter "W" alone consisted of one thousand lights.[111] Chicagoans mounted an electric light picture of Dewey's flagship, *Olympia*, on scaffolding at the corner of State Street and Madison. According to a contemporary description, "The ship itself was outlined by 720 eight–candle power lamps, 200 red-bulb and 150 blue-bulb lamps being used. A 10,000–candle power, 35-ampere searchlight was placed in the pilothouse of the ship. Portraits of Dewey and McKinley were outlined by incandescent lamps."[112]

Community holidays were also a significant light opportunity. Christmas in New York in 1894 was an occasion for the current to promote seasonal cheer. According to *Electrical World,* "On house tops, on store fronts, in theatres, in the church, the glow of the incandescent lamp is to be seen, spelling out names and legends, or forming queer devices and decorations."[113] One special holiday exhibit was a colossal figure of Santa Claus with lighted eyes and coat buttons, surrounded by illuminated Christmas trees, and leaning on a balcony in Madison Square Garden, to give children "a pleasant foretaste" of the sights of the toy fair inside.

Spectacles of the Future

Upon such experiments and spectacles, grand future schemes of communication by electric light were erected with imaginative flourish. The most fanciful were proposals to inscribe the night skies with powerful beams visible to all the inhabitants of the surrounding countryside. Appearing in many variations, this proposal was a plausible extrapolation from existing technological achievements. It extended the familiar principle of the magic lantern and newer applications of electric light technology, including recent attempts to improve the reliability and safety of shipping.

The fact that vessels sailing the coast could often determine the locations of towns from reflections of their night lights off overhanging clouds inspired experiments in projecting brilliant Morse code flashes overhead from naval vessels. In one experiment these flashes were decipherable at a distance of sixty miles.[114] In another, an astonished crowd filled the streets in the vicinity of the Siemens-Halske telegraph factory in Berlin, where a searchlight strong enough to illuminate handwriting at the distance of a mile was aimed at the sky. By means of a large mirror, signals placed in front of the light "were repeated, of course on a gigantic scale, on the clouds."[115]

Other experiments attempted to implement optical telegraphy with luminous hot air balloons. An early French effort was made in 1881 by aeronauts who enclosed a spiral of platinum in a glass vial and kept it red-hot with current from two small batteries. In 1887 the citizens of Edinburgh observed "an unusual light" in motion in the mist above Castlehill, which excited their "wonder and curiosity."[116] This was a signaling experiment with half a dozen incandescent lamps, each mounted in a wire cage and inserted into a large balloon. Military authorities in Belgium attached six lamps to the bottom of a balloon and connected them by wire to batteries on the ground in 1888. The next year in Antwerp, lights flashing in Morse sequence from a balloon three hundred feet above the ground were visible for several miles.[117] French aeronauts using a lamp and reflector exchanged light signals with telegraph operators stationed high on the third platform of the Eiffel Tower in 1890.

But it was the prospect of illuminated messages on the slate of the heavens that most fascinated experts and laymen. "Imagine the effect," speculated the *Electrical Review*, "if a million people saw in gigantic characters across the clouds such words as 'BEWARE OF PROTECTION' and 'FREE TRADE LEADS TO H--L!'" The writing, it added, could be made to appear in letters of a fiery color.[118] According to one electrical expert, "You could have dissolving figures on the clouds, giants fighting each other in the sky, for instance, or put up election figures that can be read *twenty miles away.*"[119]

In 1892 *Answers* reported that the manager of a large firm known to one of its writers received frequent suggestions for advertising gimmicks. "Often . . . in various totally unworkable forms, has the idea of casting a reflection of our 'ad.' on clouds, à la the spectre of the Brocken, been mooted to us."[120] Since projects of this kind were usually undertaken for commercial ends, the popular term for celestial projection was "advertising on the clouds." As early as 1889, an American inventor claimed to be negotiating with several firms that wished to "display their cards" in the sky.[121] Several years later, a British experimenter was said to have successfully projected the letters "BUF" upon the clouds, but his target was too small to accommodate the rest of the message: "FALO BILL'S WILD WEST." Even portraits had been "placed" on the clouds, the account continued, "though the report does not say how great the resemblance was."[122] Advertising on the clouds, explained E. H. Johnson, was "simply a stereopticon on a large scale" that required a light sufficiently powerful to reach the firmament, and a method for focusing diffuse light on a "cloud canvas" constantly shifting its distance from the earth.[123] Lecturing in Medford, Massachusetts, on electric light projection, Amos Dolbear reported in 1896 that "one may now see, in New York City, upon the dark background of the clouds, how many advertisements he can get for a dollar alternating with the saintly countenances of Lydia Pinkham or the Old Judge—while the lightning plays about the pictures."[124]

While conducting searchlight experiments on Mount Washington over a period of months, a General Electric engineer named L. H. Rogers received a stream of letters from a viewing public as far away as 140 miles. The Morse code flashes that Rogers projected on the heavens were easily read in Portland, 85 miles away. He marveled to think that "hundreds of thousands of eyes were centered on that one single spot, waiting for the flash or wink of the 'great luminous eye'— which had recently come among them."[125] Reflecting on the Mount Washington experiment a year later, Amos Dolbear imagined the day when great stencils of tin and iron would project "advertising-sheets" on the clouds with letters more than a hundred feet long, legible a mile or more away, and when weather forecasts "can be given by a series of flashes of long and short duration, constituting a code of signals, and thus the probable changes in the weather announced."[126] In Britain, electricians at Earl's Court were said to have outdone the Mount Washington experiments "not only in throwing the distinct forms of gigantic letters upon the clouds, but they have even made the well known features of Mr. Gladstone appear in ghostly outlines in the heavens."[127] The eventual outcome of the Rogers experiment in casting "legible lines" on the clouds was an electrically powered monster magic lantern that was mounted atop Joseph Pulitzer's *World* building, then the tallest building in New York, for cloud projection and advertising. It had an illumination of 1.5 million candlepower and weighed well over three thousand pounds. An eight-inch lens projected stencil-plate slides of figures, words, and advertisements upon the clouds or, on clear nights, nearby buildings.[128]

During the New York gubernatorial election of 1891, the *World* used its lantern with a code of long and short flashes to report on the contest between Jacob S. Fassett and Roswell P. Flower. The progress of the race was thus easily followed by residents in Long Island and New Jersey. Nearby, the *New York Herald* used the powerful searchlight at Madison Square Garden to project a "brilliant pencil of light" to the west when Fassett seemed to be winning. If Flower were ahead, "the eastern heavens were illumined." The *Boston Post* issued its own "sky edition" with information about returns from Massachusetts, New York, and Ohio. Learning that a favored candidate had won, it "stabbed the zenith with a perpendicular ray of light for ten minutes."[129]

The following year, the *Herald* and the *World* broadcast presidential election returns with additional searchlights "to give the hungry thousands news of the great political battle."[130] This time the dome of the *World* building was equipped with vertical bands of colored light coded for different meanings. When the *Herald*'s Madison Square Garden searchlight shone southward,

> it meant that New York had gone for Cleveland, and when the beam illuminated Brooklyn it was thought that Cleveland was certainly elected. If New York State had gone for Harrison, Harlem would have been illuminated, and if Harrison had been reasonably sure of carrying the country, the darkness of New Jersey would have been pierced by the penetrating pencil of light.

Other newspapers, including the Chicago *Herald*, also used searchlights to report election returns.[131]

Objections to the vulgarity of marketing messages on the sky were frequent. *Answers* called the possibility of sky signs "The Newest Horror" in 1892:

> You will be able to advertise your wares in letters one hundred feet long on the skies, so that they will be visible over a dozen counties. As if this truly awful prospect were not enough, we are told that these sky-signs can be made luminous, so that they will blaze away all night! A poet, in one of his rhapsodies, said that he would like to snatch a burning pine from its Norway mountains and write with it the name of "Agnes" in letters of fire on the skies.
>
> But he would probably not have cared to adorn the firmament with a blazing description of somebody's patent trouser-stretcher, or a glowing picture, as large as Bedford Square, of a lady viewing the latest thing in corsets.[132]

An English newspaper decried "celestial advertising" as the means by which "the clouds are to be turned into hideous and gigantic hoardings. This awful invention deprives us of the last open space in the world on which the weary eye might rest in peace without being agonized by the glaring monstrosities wherewith the modern tradesman seeks to commend his wares."[133]

If the sky was a logical surface upon which to reflect messages for the millions, so was the moon. *Science Siftings* reported in 1895 that an American named Hawkins planned to send a flashlight message from London to New York via the moon, using a gigantic heliograph reflector to catch the sun's rays and cast them on the moon's surface. Hawkins had conceived the intellectual principles of satellite relay using the only earth satellite available in 1895. The value of his plan, he announced, lay in covering long distances,

> but electricity would be required for local distribution from the receiving stations. If a flash of sufficient strength could be thrown upon the moon to be visible to the naked eye, every man, woman and child in all the world within its range could read its messages, as the code is simple and can be quickly committed to memory.[134]

Discussions such as this often took for granted a technologically driven transformation in the scale of the audience for new modes of communication. Both fantasies of communication with intelligent aliens at interplanetary distances and fantasies of global communication titillated the collective capacity for imagining the social limits of new media. They likewise instructed collective imagination to explore the possibility of dramatic shifts in the social order in an age of communications transformed. Since laying a cable to outer space seemed unlikely, signaling schemes to strike up a wireless conversation with extraterrestrial beings received wide publicity. One suggestion was offered by Amos Dolbear, who proposed that a powerful searchlight such as was exhibited at the Columbian fair, having the power of millions of candles, can be directed in a dense beam and can be made intermittent; signals can thus be sent the same as from the tops of mountains. Once out of the air there would be no loss from absorption and the beam would speed on, reaching Mars in about four minutes when it is nearest us.[135]

He added in his regular *Cosmopolitan* column, "The Progress of Science," in 1893 that communication with Mars would be possible "if it should chance to be peopled with intelligences as well equipped with lights and telescopes as we are."[136]

Other schemes proposed semaphoric arrangements of giant lights flashing in Morse code sequence. The *Live Wire,* a Munsey dime monthly whose title bore witness to the popular association between excitement, novelty, and electricity, reported that Sir Francis Galton had proposed to construct heliograph mirrors seventy-five feet by forty-five feet to flash a regularly pulsing ray of sunlight to Mars.[137] The founder of the first electrical engineering curriculum in the United States, physicist Charles Cross, thought that the beam of a powerful electric light might be gathered and concentrated at a single point by huge parabolic reflectors. A plan to use some of the noblest monuments to the history of the race to convey visible signs of earthly civilization suggested that "incandescent lights be strung over the sides of the Great Pyramid, and thus it be made a great square of light. When it was pointed out how inadequate this would be, the proposer replied by saying, 'Then illuminate all the pyramids.' "[138]

The science-fiction novelist Camille Flammarion offered a plan to group immensely powerful lights in the pattern of a familiar constellation like the Big Dipper for extraterrestrial observers. Its dimensions would require lights at Bordeaux, Marseille, Strasbourg, Paris, Amsterdam, Copenhagen, and Stockholm. "But no one has yet been found to build seven lights each of about three billion candlepower," explained the *Live Wire.*[139] The popular Astronomer Royal of Ireland, Sir Robert Ball, hoped to dispense terrestrial nationalism to a celestial audience with "an intensely luminous flag as large as Ireland" for Martians to see, if any there were.[140] A plan to work out an interesting geometrical problem in lights for the amusement of galactically remote viewers was impractical, the *Live Wire* concluded, because the lines of every figure would have to be at least fifty miles wide "and made of solid light." (A similar proposal was made to build a Pythagorean triangle in the Sahara Desert by planting palm tree forests, to stand in green contrast to the white glare of the sand.[141] Great Britain's Assistant Astronomer Royal, E. W. Maunder, speculated that "if ten million arc-lights, each of one hundred thousand candle-power, were set up on Mars, we might see a dot."[142] Intergalactic light signaling was a sufficiently current topic to elicit comments even from those who doubted its efficacy. The British *Spectator* greeted the announcement of William Preece's experiments in wireless telegraphy with the observation that interplanetary communication by wireless telegraphy was a better plan than "any scheme for making gigantic electric flashes, which the Martials, assuming them to exist, might mistake for lightning of an unusual degree of force."[143]

Each of these otherwise highly various images of mass audiences viewing electric light messages in the night sky assumed that such audiences naturally belonged outdoors, and that familiar nighttime social gatherings illuminated for dramatic effect as well as utility would be expanded on a grand scale in the future. As things have turned out, contemporary mass audiences congregate mostly indoors and not together. But other elements of early illuminated gatherings do point directly to one of our most familiar modern public spectacles, broadcast entertainment. Television's inheritance from the electric light is both technological and social. The original electronic effect, the so-called Edison effect, though poorly understood at the time of its discovery, was created in an electric lamp. The development of electronic tubes and transistors out of this puzzle in a light bulb eventually helped make many face-to-face public gatherings superfluous as families and individuals retreated indoors to well-lighted living rooms to watch on television the descendants of the public spectacles that had once entertained communities in the town square. The television special is a still identifiable heir of this genre, using brightly colored lights in striking patterns and images to create visual excitement and drama.

Because communication at a distance was mostly implemented in other forms, our cultural memories no longer recall predictions in nineteenth-century voices that twentieth-century media might include messages splashed across the sky by searchlight or projected on walls by banks of electric lights. The nineteenth-century conviction that important twentieth-century mass media would look like nothing so much as nineteenth-century electric lights writ large betrays the tendency of every age to read the future as a fancier version of the present. For late-nineteenth-century observers, the electric light was a far more likely mass medium than any point-to-point invention such as the telephone or even wireless, which the nineteenth century regarded as a vastly imperfect point-to-point medium. Most people made the acquaintance of electric light through its decoration of outdoor spectacles that candles and bonfires had once illuminated, and in new spectacles such as the sports event after dark, the nighttime public amusement park, and the electrical exposition. Electric spectacles were the first dramatic ritual event created for the public by electrical professionals. These outdoor events were complemented by dramatically lighted indoor occasions for smaller gatherings, at first as a mark of conspicuous consumption, in time as a matter of routine.

Transformative patterns of this kind are common in technological innovation. Apparatus intended to streamline, simplify, or otherwise enhance the conduct of familiar social routines may so reorganize them that they become new events. The lines of their evolution remain, however. Ordinarily, we think of wireless telegraphy, cinema, and telephony as the direct ancestors of mass broadcasting, but this genealogy overlooks the role of electric light in the social construction of twentieth-century mass media. The communicative capacity of electric light survives today in illuminated signs, but its most important contribution to modern mass communication was to a vocabulary of popular forms in mass entertainment spectacles and to the reorganization of traditional audiences. In that sense, the glittering television special is as much the fruit of electric light as of any other invention.

5

Annihilating Space, Time, and Difference

Experiments in Cultural Homogenization

> The little Alpena [Michigan] *Echo* cut off its daily telegraph service because it could not tell why the great head in the telegraph company caused it to be sent a full account of a flood in Shanghai, a massacre in Calcutta, a monkey dance in Singapore, a sailor fight in Bombay, hard frosts in Siberia, a missionary banquet in Madagascar, the price of kangaroo leather in Borneo, and a lot of nice cheerful news from the Archipelagoes, and not a line about the Muskegon fire.
>
> —Detroit News, 1891

The most admired feats of the telephone, cinema, electric light, phonograph, and wireless were their wonderful abilities to extend messages effortlessly and instantaneously across time and space and to reproduce live sounds and images without any loss of content, at least by the standards of the day. Experts and publics agreed on the brilliance of this achievement. But wherever these extraordinarily sensitive new nerve nets extended, there was little genuine sense of cultural encounter and exchange. In electrical publications of the late nineteenth century, newly accessible lands and people were seldom cherished for any cross-cultural opportunities they offered, except abstractly. Concretely, they appeared as islands of cultural anomaly that new techniques of communication made available for absorption into the mainstream. Those who controlled the new electrical technologies not infrequently dismissed vastly different cultures as deficient by civilized standards, lacking even the capacity for meaningful communication.

What late-nineteenth-century writers in expert technical and popular scientific journals practiced was a species of cognitive imperialism. Theirs were visions of a globe efficiently administered by Anglo-Saxon technology, perhaps with exotic holidays, occasions, and decorations in dress and architecture, perhaps filled with more items and devices than any single person could imagine, but certainly not a world to disturb the fundamental idea of a single best cultural order. What these writers hoped to extend without challenge were self-conceptions that confirmed their dreams of being comfortably at home and perfectly in control of a world at their electric fingertips. Even when, in the utopian manner, their declared goal was to turn the status quo upside down in pursuit of a better world, few of their schemes failed to reconstitute familiar social orders and frameworks of interpretation. Only the scale of the community in which they imagined themselves as participants had changed.

Changes in the functional capabilities of new media of communication were a matter of interested discussion by electrical scientists, engineers, entrepreneurs, and camp followers. Suggestions that the future of these devices lay in the organization of public intelligence systems to promote cultural harmony and perfection by displaying it to one and all were sympathetically received. With new communications techniques, the idealized world of technologists would be extended automatically to the less fortunate periphery—less fortunate *because* it was at the periphery. This sentiment was widely held. "Any device that enlarges one's environment and makes the rest of the world one's neighbors," telephone inventor Amos Dolbear once remarked, "is an efficient mechanical missionary of civilization and helps to save the world from insularity where barbarism hides."[1] An early prophet of transoceanic telegraphic communication, Alonzo Jackman, offered a more explicit cultural vision of the salvation of the world through instantaneous long-distance communication in a "new era" of evangelism. "Heathenism would be entombed, and the whole earth would be illuminated with the glorious light of Christianity."[2]

In a discussion of the future of wireless in 1904, Nikola Tesla was confident that wireless would be "very efficient in enlightening the masses, particularly in still uncivilized countries and less accessible regions, and that it will add materially to general safety, comfort and convenience, and maintenance of peaceful relations."[3] The world society that would arise spontaneously in the absence of barriers to communication would resist autonomous minorities. "If every country, even the smallest," Tesla had observed several years earlier, "could surround itself with a wall absolutely impenetrable, and could defy the rest of the world, a state of things would surely be brought on which would be extremely unfavorable to human progress. It is by abolishing all the barriers which separate nations and countries that civilization is best furthered."[4] Those thought to offer the greatest obstacle to human progress were at the periphery of the cultural nerve net, never transacting as equals with the center, and earning a hearing only if they were already appropriately submissive in the fantasies of those seeking to dominate them.

"The means of communication which are the signs of the highest forms of civilization," the president of the Canadian telegraph system, Erastus Wiman, had explained to the New York Electric Club in 1899, "are the most perfect by aid of electricity simply because they are instantaneous. There is no competition against instantaneousness."[5] Unavoidable engagement, enforced by new technologies of communication and benignly labeled "quick sympathy," would end jealous nationalism, in the view of *Science Siftings* and many other journals. Instantaneous electric communication augured a universal language, usually thought to be English, and global harmony.[6] This distinctly Anglophile solution reflected a conviction that the provincialism of English-speaking peoples was the sensibility of the world. Similar accounts described land and telephonic lines covering "most civilized countries" with a network of wires, and the rest of the world as "within a few seconds' or minutes' communication with London" or New York, always orienting the rest of the world to its own center, never imagining a world of equidistant points in which every place was equal to every other.[7] The vastly extended eyes- and ears-to-be of new machines of communication anticipated few cultural puzzles to unravel, and showed their inventors only the most reassuringly echoic and potent images of themselves.

The Self-Centered Universe of Media

Communication and Cultural Difference

Two recurring themes about new media addressed the issue of social boundaries in transition, though with a certain indirectness. One attempted to predict technical advances in media of the future, and the other speculated about what the effects of these advances might be. The devices that social imagination constructed and then reacted to sometimes actually existed, but just as often were entirely imaginary.

Many are unrealized fantasies still. Even among observers with scientific expertise, few in the volatile atmosphere of the late nineteenth
century could be certain what wild fantasies might already have been
translated into technical realities.

New media were recognized as new because they exhibited certain
features. New media "truthfully transferred," in Thomas Edison's words,
an increasing number of the auditory, visual, and kinesthetic details
of the occasion of communication.[8] New media also addressed expanding audiences, whether across time or space, and whether the audience attended all at the same time or in varying combinations, as in
a telephone network. The more any medium triumphed over distance,
time, and embodied presence, the more exciting it was, and the more
it seemed to tread the path of the future. Such achievements were often
imagined in great detail. And always, new media were thought to hail
the dawning of complete cross-cultural understanding, since contact
with other cultures would reveal people like those at home. Only physical barriers between cultures were acknowledged. When these were
overcome, appreciation and friendliness would reign. If contact did not
reveal people exactly like those at home, it would reveal people anxious to learn from those at home. Never would it reveal things that
those at home needed to learn. Assumptions like this required their
authors to position themselves at the moral center of the universe, and
they did. They were convinced that it belonged to them on the strength
of their technological achievements.

Ambitious schemes for communicating with all the earth and even
beyond the stars, schemes that aimed to overcome the limitations of
distance and wire, were often modeled on familiar machines such as
watches, compasses, telephones, and telegraphs. In 1880 Harvard professor John Trowbridge proposed to the American Academy of Arts
and Sciences a scheme for wireless transmission extrapolated from his
own induction experiments with time signals. Trowbridge had detected
electric time-keeping pulses transmitted from Cambridge Observatory
to Boston in a six-hundred-foot-long grounded wire located parallel to
and a mile away from the signal wire. From this success, he dared to
imagine wireless induction powerful enough to leap the ocean:

> Powerful dynamo electric machines could be placed at some point in
> Nova Scotia, having one end of their circuit grounded near them and
> the other end grounded in Florida, the conducting wire consisting of a
> wire of great conductivity and being carefully insulated from the earth
> except at the two grounds. By exploring the coast of France, two points
> on two surfaces not at the same potential could be found, and by means

of a telephone of low resistance the Morse signals sent from Nova Scotia to Florida would be heard in France.[9]

A similar plan had been advanced by J. W. Wilkins in *The Mining Journal: Railway and Commercial Gazette* in the earliest days of partnership between the telegraph and the railroad. He proposed to lay parallel wires, each twenty miles long, on either side of the English Channel.[10] He envisioned an expanded telegraphic receiver and transmitter operating on wireless induction principles. According to the most detailed of his proposals:

> Suppose an electric circuit in America, one terminal say in Greenland, the other in Brazil. A dynamo and a telegraph key would enable one to send an intermittent current into the earth at those terminal places. At the Greenland end the current would spread out to a tremendous extent, some of it coming through North America, some through the ocean and some going to Europe, to cross the ocean again between Africa and South America. If a telephone circuit were placed on the eastern shore of the Atlantic, with its ends buried in Norway and in Africa, every Morse signal produced on this side would be heard on the European side, though the breadth of the ocean were between the wires.[11]

The same reasoning that extrapolated large technical systems from homely and familiar devices designed to work on a smaller scale was extended to beliefs about the nature of the world culture in which messages would be exchanged. Many writers assumed that messages that made sense at home, and the assumptions on which they were based—especially acceptable and proper procedures of hospitality, courtesy, and reciprocity—would make sense all over the world. If they did not, the flaw would be always with the Other, whose inferiority could be demonstrated if proof were needed.

The capacity to reach out to the Other seemed rarely to involve any obligation to behave as a guest in the Other's domain, to learn or appreciate the Other's customs, to speak his language, to share his victories and disappointments, or to change as a result of any encounter with him. For their part, peripheral Others were expected to do all these things, to communicate on terms provided by the center, and to converse with the representatives of European civilization without saying much back in the course of that conversation about their own unique cultures. These conditions of communicative exchange were reminiscent of the relationship between male electricians and that most exotic of genders, the female.

A feature story in the *Reading* (Pennsylvania) *Herald* of 1889 describing the novelty of long-distance telephone service between New York and Boston may be taken as a prototypical expression of excitement and pride in the expanding social universe that new media would create. This tale was an odyssey of partners in heroic magic: the electric current and the power of a word. This was not a secret password guarded by an elite, but an expression of friendship, an American Everyman word taken to be unpretentiously and universally comprehensible. The current that carried it was no royal-adventure fleet battling dangerous seas and alien perils, no curious wanderer stopped short in amazement by the unexpected or the transforming. The universe of its travels was the familiar expanse of American urban civilization and its immediate surround. Its journey began in the central office of the New York Telephone Company, "crammed full of electricity from basement to roof," among hundreds of "people who deal in tame lightning," where man was neither adversary to nature nor awestruck by its creations, but master of its miracles:

> That Hello! took advantage of its opportunities and travelled. It went down through the desk, down through the floor into the basement of the building, then out into an underground conduit, rushing along under all the turmoil and rush of New York City, then up the Hudson, taking a squint at the Palisades, past Yonkers and Tarrytown and Sleepy Hollow, then out into the land of the wooden netmegs [sic], to and through New Haven, Hartford, Providence, Newport, on to Boston.
>
> It crossed rivers and mountains, traversed the course of fertile valleys and past busy factories, noisy with the whirr of a thousand machines. It went through lonely forests, past places where they used to burn witches, and scenes familiar to the Pilgrim Fathers. It heard the music of the sea, it saw the homes of the rich and the poor and it caught a glimpse of Bunker Hill monument before it plumped into the city of baked beans and reached its destination in the ear drum of a man seated in a high building there. . . .
>
> In about one millionth of the time it takes to say Jack Robinson, it was there. It had turned a thousand curves, it had climbed up and slid down a hundred hills, and yet it came in at the finish fresh as a daisy on a dewy June morning. It was as if by a miracle the speaker had suddenly stretched his neck from New York to Boston and spoken gently into the listener's ear. It beat all to smash all the old incantations of Merlin and the magii [sic] of Munchhausen, Jules Verne, or Haggard. In plain United States, it was the long distance telephone.[12]

The itinerary of the long-distance hello evoked reassuringly homogeneous images of cultural identity. The long-distance telephone was a sentimental pioneer leading the American past into the future.

Nevertheless, the standard-bearer of democratic universalism was a provincial at heart. It embraced the obedient ideology of industriousness. The differences it applauded were those of the physical landscape alone. Experience at a distance consisted in recognizing friends, not in becoming acquainted with strangers, and in confirming the idealized pattern of events in industrial societies.

Perhaps the clearest account of what was generally hoped for in cross-cultural contact appeared in stories about inventions that did not exist, and about communication with cultures that did not exist, concerning which, therefore, unfettered fantasy could run free. This was how speculation went about the "telectroscope," a popular but entirely imaginary invention of the late nineteenth century "by which actual scenes are made visible to people hundreds of miles away from the spot." The telectroscope was the perfect medium for the late nineteenth century to speculate about, one that had to be created, to paraphrase an old line, because it didn't exist. For reaching a vastly expanded audience, for reconstructing every dimension of experience in faithful detail, and for freeing communication from every physical constraint, the telectroscope was regarded in many quarters as the final triumph of electrical communication. Though continuously disappointing prophets who awaited its appearance at one after another of the large international fairs of the late nineteenth century, the telectroscope was described even by the *Electrical Review* as the most promising project yet of the country's most famous inventor for the most spectacular of fairs, the Chicago World's Fair. It was said that when it appeared, Edison's latest and most remarkable invention would "increase the range of vision by hundreds of miles, so that, for instance, a man in New York could see the features of his friend in Boston with as much ease as he could see the performance on the stage."[13]

A Society of Arts lecture in 1894 by the chief engineer of the post office on the subject of electric signals to Mars was still more daring. "If any of the planets be peopled with beings like ourselves," rhapsodized William H. Preece in a burst of optimism on behalf of an intergalactic sympathy for which there was no evidence of any kind, "having the gift of language and the knowledge to adapt the great forces of Nature to their wants"—specifically, "if they could oscillate immense stores of electrical energy to and fro in telegraphic order"—then "it would be possible for us to communicate by telephone with the people of Mars."[14] Occasional reports of possible extraterrestrial attempts to contact Earth always featured space aliens signaling in codes familiar to English-speaking peoples. An astronomer's report in 1892 that Mars was signaling Earth in enormous blue capital letters was widely carried in the popular press and humorously noticed in the scientific press.[15] A 1907 source reported that undecipherable messages often came in by wireless "between 12 and 1 a.m.," and were finally determined to be the three dots that meant *s* in Morse code.[16] Where it did not exist, and could not exist, cross-cultural, even cross-galactic communication was premised on a model of instant mutual sympathy. In electrical ideology the barriers of incomprehensibility were time, distance, accessibility, and experiential fidelity. If any failure of understanding occurred, it could not be the fault of the culture that had defeated the barriers to communication.

Predictably, the *experience* of contact between distant cultures met few expectations of mutual recognition. For Thomas Stevens, a British telegraph operator in Persia responsible at the most personal level for bringing the kinship of humanity closer to fruition, the telegraph was not a device to facilitate contact with a remarkably different and fascinating culture, but an intellectual and spiritual restorative in a cultural as well as physical desert. "How companionable it was, that bit of civilization in a barbarous country, only those who have been similarly placed know." A system of rewards and punishments assigned British operators to favored stations in Tehran, Isfahan, or Shiraz, or to disfavored remote stations where contact with the locals was regarded as valueless except when the operator was treated as a potentate. The telegraph represented "a narrow streak of modern civilization through all that part of Asia." Europeans as far apart as two thousand miles, who had never seen one another, were well acquainted. Many were like "the lone Englishman at the little anterior control station of Dabeed," who wished to hear from his chums at Tehran only "of the doings of the European colony there."[17] One could not talk to a culture that did not know how to talk. "It is evidently a poor place for the telephone in the land of the Arabs," mused the editors of the *Electrical Review*. "The nearest they can come to it is to throw a stone and hit a man in the back, and then ask him, as he turns around: 'Does it please Heaven to give you good health this morning?'"[18]

At the romantic remove of prophecy, global conversations were always universally intelligible. "Some years ago," the *Electrical Review* informed its readers, *Athenaeum* editor W. Hepworth Dixon had made a provocative prediction:

> Two armies clash, a victory is gained, an empire rises, and a second
> empire falls. Events occur in a few hours which change the flow and

custom of the world. A crash, an onset and a rout. Napoleon a prisoner, Wilhelm is on his way to Versailles. The political and military center of Europe is transferred from Paris to Berlin. These things are done in a dozen hours, and in another dozen men are talking in their breathless haste and fever of these great events, not only in Paris and Berlin, but in the mosques of Cairo and in the streets of Arkangel, in the bazars of Calcutta and on the quays of Rio, by the falls at Ottawa, in the market place of San Francisco and in the shops at Sidney [sic], within a day the news is told, and at the same instant of time every human heart is quivering with the shock of these great events. That is drama. All the corners of the earth are joined, kindled, fused. Just as in a theater you speak directly face to face with five or six hundred persons, so that wave of merriment, crying with the same pang of emotion, so the poetical telegraph speaks to the whole world—now become a theater—brings us joy and sorrow, exultation and remorse, and every kind and race of man.[19]

In this extract, it is important to note, only the audience is in all corners of the world. The stage play is entirely in Europe. While the corners of the earth were flatteringly "breathless" in attending to the apparent center of human drama, that center was oblivious to the possibility of uncertainty, misunderstanding, or hostility in encounters between cultures with different logics of experience, or indeed to the possibility that interest might ever lie elsewhere.

Assessing the "moral influence" of the telegraph in the wake of international reaction to the assassination of President William Garfield in 1881, *Scientific American* concluded that the "kinship of humanity" had been inconceivable before the telegraph. The tragedy had mesmerized the entire world, according to the editors. How the world had developed its fascination with American political life in the absence of previous opportunities for regular, rapid communication did not seem to be an interesting point.

It was the touch of the telegraph key, a favorable opportunity being presented, that welded human sympathy and made possible its manifestation in a common, universal, simultaneous heart throb.

We have just seen the civilized world gathered as one family around a common sick bed, hope and fear alternately fluctuating in unison the world over as hopeful or alarming bulletins passed with electric pulsations over the continents and under the seas. And at last, on the same day, the nations stand in sympathetic mourning: a spectacle unequalled in history; a spectacle impossible on so grand a scale before, and indicative of a day when science shall have so blended, interwoven, and unified human thoughts and interests that the feeling of universal kinship shall be, not a spasmodic outburst of occasional emotion, but constant and controlling.[20]

Besides offering a long-deprived humanity access to its putative emotional center, another advantage of new media was to render inhabitants of that center ever less physically and morally obliged to stir from it. With a sufficiently powerful illusion of enriched experience, the culture of the center could with a clear conscience further reduce the already narrow boundaries of its experience. "Has not this Galician genius done away with the necessity of visitors actually going to Paris in 1900?" mused *Lightning* about the telectroscope Jan Sczepanik claimed to have invented for the Paris Exposition.[21] "Possibly the time will come when so far as seeing objects are concerned, one can make a tour of Europe without going out of his own house," surmised the *Electrical Review* in 1889.[22] Faintly echoing the polar visions of Famianus Strada, Edward Bellamy portrayed a similar future for entertainment in his novel *Equality:*

> You stay at home and send your eyes and ears abroad for you. Wherever the electric connection is carried—and there need be no human habitation how ever remote from social centers, be it the mid-air balloon or mid-ocean float of the weather watchman, or the ice-crushed hut of the polar observer, where it may not reach—it is possible in slippers and dressing gown for the dweller to take his choice of the public entertainments given that day in every city of the earth.[23]

With a like preference for the passive safety of routine over the novel risk of activity, *Comfort* predicted that soon "We may hope to have the movements, the actual happenings of the world, as they are transpiring, brought to our firesides in the form of pictures. Then we shall no longer need to go in person to witness a ball game, or to the inauguration of a president."[24] By such strategies, fantasies of expanded experience could be entertained within the steadily contracting circle of the physical, emotional, intellectual, and moral fireside. By an imaginative sleight of hand that denied the paradox it created, more and more of the world could be "experienced" in an increasingly familiar and restricted space and time, where rude cultural differences would not intrude on expectation and routine, and cultural difference itself could be regarded as illusory or beneath serious notice. One's own family and neighborhood would then be the stable center of the universe—beyond it would be margin and chaos.

Nor surprisingly, the prospect of media that made senders and receivers proximate and seemed to eliminate many of the barriers that kept them safely separated excited profound xenophobic anxiety. Professor Alonzo Jackman mulled over the risks of physical connection by wire to those who were diseased, or who might in some other way be dangerous in a telegraphically linked world. He concluded that technology was as useful for prophylaxis as for making contact in the first place. In a letter to the *Woodstock* (Vermont) *Mercury* in 1846, Jackman proposed to connect American and European capitals with a Grand Submarine and Overland Magnetic Telegraph of parallel undersea iron cables carrying messages by electrical induction. The benefit of the arrangement would be that "All the inhabitants of the earth would be brought into one intellectual neighborhood and be at the same time perfectly freed from those contaminations which might under other circumstances be received."[25]

The utopian retreat from variety in experience was not limited to contact with remote cultures. With long-distance communication, those who were suspect and unwelcome even in one's neighborhood could be banished in the name of progress. In a short *Cosmopolitan* fantasy by Julian Hawthorne about the world of 1993, the crowded cities of late Victorian landscapes were rationally reshaped into communities of homogeneous culture, race, language, and self-selected association. Cities would cease to be colorfully diverse containers of unpredictable life. Industrial concentration would exist only at four efficient points on the continent. Families would be dispersed throughout the country on uniform lots of land on which they would live economically self-sufficient, separate existences. Five or six times a year they would gather at special cultural centers with theatres, churches, museums, and pleasure gardens. These institutions of organized rather than spontaneous interaction would provide the only unplanned occasions for face-to-face exchange.

This regimented avoidance of uncertainty in social relations was said by Hawthorne's fictional guide to foster a healthy psychic isolation that led men to "comparatively an interior, and therefore a more real and absorbing life. For the first time in history we have a real human society."[26] In one of the stock phrases of the late nineteenth century, flight, telephones, and telegraphs had "brought every individual of the nation into immediate and effortless communication with every other," and had made it possible for compatible individuals everywhere to discover one another.

> Thus, each family lives in the midst of a circle of families comprising
> those who are most nearly at one with it in sentiment and quality, and

the intercourse of this group is mainly confined to itself. There is between them perfect and intimate friendship and confidence, and you will
easily understand that they must realize the true ideal of society.[27]

Political geography had been "practically obliterated," free trade was
universal, "there were no longer any foreigners" anywhere. Global life
was so well integrated that everywhere there was the "gradual adoption
of a common language." Ultimate proof of all:

> [T]oday the inhabitants of this planet are rapidly approximating to the
> state of a homogeneous people, all of whose social, political and com-
> mercial interests are identical. Owing to the unlimited facilities of in-
> tercommunication, they are almost as closely united as the members of
> a family; and you might travel round the globe, and find little in the
> life, manners and even personal appearance of the inhabitants to remind
> you that you were remote from your own birthplace.[28]

Accepting a challenge from his host, and speaking as the straight man
of this narrative, Julian Hawthorne pegged him for a New Englander
of Welsh descent. The pleased host claimed an ancestry of "unadul-
terated Esquimaux." All peoples of the earth had become nearly in-
distinguishable. In particular, "a change has evidently taken place in
the interior physical constitution of the dark races, causing them to
tend both in form and hue towards the Caucasian standard."[29] Life in
utopia, always the Caucasian standard.

The Uses of Looking Backward

When the printing press was introduced in Europe, Elizabeth Eisenstein has argued, its effect turned out to be different from anything the
Catholic Church might have hoped for.[30] The power of the press to
reproduce copies quickly and accurately led the Church's theologians
to believe that questions concerning the true text of the Bible could be
definitively settled at last. By reproducing manuscript fragments in
printed form for convenient comparison, the Church's best scholars
hoped to arrive at an authoritative scriptural text. Instead, Biblical
scholarship in the age of the printing press opened permanent rifts in
Western religious culture. Comparing printed manuscripts led scholars
to the conclusion that a unitary Word had not, after all, been corrupted
by the vagaries of copying and preservation over the centuries. The
Bible they discovered was a collection of manuscripts produced at different times and in different traditions, consolidated perhaps in the historical imagination of men, but not of God.

This historical lesson provides a key for understanding popular reaction to another preservative medium within the framework of social expectation outlined here. Like the electric media of spatial extension that promised a way to impose order from the center on the periphery, the preservative capacity of the phonograph offered a means of controlling past and future. Discussions about the phonograph approached it as an invention of great social importance. As usual, praise for its artifactual capabilities as a recorder of experiential fidelity concealed a more complicated social agenda. The phonograph provided a focus for concern about cultural stability in an era of rapid change. More, it offered a chance to lay the foundation for an ideal future culture based on a chosen past. The past of choice was European high culture. The present age, in the words of *Scientific American,* could look "forward to a time when faculties such as Shakespeare, Newton, Mozart, Michael Angelo [sic], and other men of great genius enjoyed shall be the common inheritance of the race."[31] Many regretted that the historical absence of lifelike preservative media had long deprived mankind of a reliable record to use as a moral and intellectual reference point, a resolver of controversy in important debates. "What we have missed!" wrote the Washington correspondent of the *St. Louis Globe-Democrat* in 1888:

> Suppose we could have graphophonic communication with the year in which Plato lived and philosophized, and we could listen to his voice and hear his discourse. What a world this would be if we could take down our machine every day and grind out a love scene between Antony and Cleopatra, or a quarrel between Romeo and Juliet. . . . How interesting it would be to find graphophones in the pyramids of Egypt. . . . Now suppose there had been phonographs in the Garden of Eden, and there had been handed down to us the cylinders recording the conversation that took place from the morning of man's birth until he was driven from out of the gates of Paradise. How easy it would be for us to sit down, turn the crank and listen to Adam's courtship and to Eve's interview with the serpent. What an endless number of acclesiastical [sic] disputations would have been avoided, and how unanimously all the doctors of divinity would have long since agreed.[32]

Guarding culture was tantamount to guarding truth. The greater the value of a cultural domain, the more potent any truth-wielding instrumentality that represented it. "Had Beethoven possessed a phonograph," the *New York Evening Post* speculated,

> the musical world would not be left to the uncertainties of metronomic indications which we may interpret wrongly, and which at best are but

feeble suggestions; while Mozart, who had not even a metronome, might have saved his admirers many a squabble by giving the exact fashion in which he wished his symphonies to be played. . . . Future generations will be able to learn, if they care to know, exactly how Rubinstein "phrased" the "Emperor" concerto, or with what mannerisms Mme. Patti sang "Home Sweet Home."[33]

"We of to-day could have taken our phonos out on the back stoop in the long summer evenings and listened to the roar of the lions in Daniel's den, the sound of Nero's fiddle and the clatter of the Roman empire as she fell," wrote the *Electrical Review*.[34]

As printing had surprised those who hoped to find the true Bible, and as electric media would surprise those who awaited the flowering of global harmony, phonographic history-in-the-making had its own unanticipated consequences. *Chambers's Journal* reported a meeting of London literati who gathered on the first anniversary of Robert Browning's death to listen to the words of the master.

> A curious point in connection with the matter is that when making the original recording on the phonograph cylinder the poet in quoting some of his own had to be prompted by a bystander, for his memory failed him. This prompting, together with the apology from the poet which followed, were duly reproduced by the instrument.[35]

Here was a matter of importance in the control of culture. What would the stature of future heroes be if the human accidents of their real lives were fixed indelibly in the permanent record of the message they left behind? Preservation was a two-edged sword. As easily as it could reduce heterodoxies of interpretation and enforce right-thinking homogeneity, it could imperil cultural stability by introducing uncontrolled variety.

To the "great picture galleries" heavy with the weight of the past, the special "vivacity" of the phonograph would add "voice galleries" more weighty still, the *Spectator* of London predicted. Its writer envisioned a twentieth-century man emotionally burdened with the accumulated conflict of his ancestral history. To the unavoidable complexity of horizontal history, the push and tug of interests at war in individual circumstances, would be added the greater complexity of collective, or vertical, history. Suppose a man had a Gladstonian great-grandfather, a grandfather who had commanded troops in the war with Ireland that had led to home rule, and a father who sided with the Irish to resist the oppressions of the restored government:

imagine these ancestors addressing their descendant in all different accents of political passions to which their different situations in life had given rise, while their portraits look severely down upon him, enforcing by their expression the earnestness of their political view—would not such a man carry into life a consciousness even more hesitating and divided than even that which gives birth to our 19th century facillations? [sic]

A time would come, predicted the *Spectator*, when

it will be necessary to preach a sort of iconoclasm toward the pieties of ancestry in order to clear the way for anything like independent growth. One of the most effective of Arabian fairy tales describes how the prince who is to break the spell of the wicked magician's enchantment has to pass along a way where voices in the greatest confusion address him on every side in every accent of scorn, or ridicule, or indignation, all appearing to come from the mere stocks and stones beside the path. . . . We much doubt whether Edison's wonderful and admirable discovery, and the extensions that must follow, will not tend to bewilder the world in which our children's children live, at least as much as the outcries of the bewitched valley of rocks bewildered the hero of that eastern tale.[36]

An event that fascinated observers with its ability to bracket and defeat time, to shape history, as it were, from a time machine, was the widely publicized phonographic "Bequest to Humanity" of Cardinal Henry Edward Manning, the controversial archbishop of Westminster, in 1892. In "long-drawn, deliberate tones" the voice of the dying cardinal began: "To all who come after me: I hope that no word of mine, spoken or written will be found to have done harm to anyone after I am dead."[37] In the hands of preservative media, there could be many versions of history, all of them troublesome.

The End of Politics in a Thing-Filled World

The idea that more communication would render cultural differences meaningless was closely related to the idea that productive abundance would render politics superfluous. A useful strategy for stripping social phenomena of the power to endanger the status quo is to anchor them to safely established notions while presenting them for public consumption as revolutionary. This was the strategy that defenders of electrical progress used to fend off unwelcome challenges to the social-world-in-place from the electrical-world-in-prospect while appearing to embrace the latter world fully and fearlessly. Electricity was frequently characterized as revolutionary to compare it to the steam revolution that had preceded it. That comparison reversed the usual meaning of *revolution* as a decisive break with the past, however. The work of electricity was presented as continuing the work of the past, and the past with all its difficulties was justified as a necessary prologue to the present and the future. At the same time, these defenders claimed that electricity was unprecedentedly revolutionary because electrical prosperity would end politics, conceived as the struggle of groups over scarce resources. The social architecture of the future was detached from every suggestion of political upheaval. The introduction of electricity was seen to have *no* political consequences, no winners or losers of power, or winners called to account for abuses of power, since politics would exist no more. The ominous meaning of the term *revolutionary* was thus neatly transformed and appropriated.

Predictions that strife would cease in a world of plenty created by electrical technology were clichés breathed by the influential with conviction. For impatient experts, centuries of war and struggle testified to the failure of political efforts to solve human problems. The cycle of resentment that fueled political history could perhaps be halted only in a world of electrical abundance, where greed could not impede distributive justice. A speech by retiring president T. C. Mendenhall to the American Association for the Advancement of Science in 1890 was noteworthy for containing so many of these clichés in a brief space: "With this spark, thanks to science, the whole world is now aflame. Time and space are practically annihilated: night is turned into day; social life is almost revolutionized, and scores of things which only a few years ago would have been pronounced impossible are being accomplished daily."[38]

Considering the extremity of the proposals, all was magnanimous equanimity. Not the structural reorganization of the polis, but lives filled with more things was the implied promise of the future. The form in which that promise was often made was a modest literary device—the list. The list is a structure with great ideological flexibility. It requires no internal logical justification or any other explicit criteria of selection and evaluation. Lists of electrical virtuosity communicated and commended an enlarged perception of things and things possible. Perhaps lists impressed those who read or heard them with visions of an abundance in the wake of which nothing else seemed necessary to consider. In a world of inexhaustible novelty, concern about the structural distribution of electrical beneficence seemed not

only less interesting than the limitless fact of its existence, but frankly
superfluous. A typical list was offered in opening remarks to the annual convention of the National Electric Light Association in New York
in 1896 by its president:

> What in the days of our childhood was scarcely more than a toy—at
> best the interesting, the mysterious phenomenon of the scientist's lab-
> oratory—is in these closing years of the century the mightiest agent
> known to man. By its means the news of the world is gathered from
> its four corners in less than a second's time. . . . Through its agency
> the softest whisper of the human voice is transmitted a thousand lea-
> gues, or recorded and preserved for the ears of future generations; the
> divisions of time are so integrated as to enable the eye to follow in
> reproduction the continuity of the most rapid motion; night is turned
> into day, darkness into light; the waste forces of nature are harnessed
> and wafted like spirits, unseen and instantaneously, over mountains and
> rivers, miles upon miles, to turn the busy wheels of distant industry;
> the hidden secrets of nature are laid bare by the ray that pierces dense
> matter with the ease of a shaft of sunlight traveling through thin air.[39]

A less elegantly phrased but no less tantalizing list appeared in
Science Siftings in 1894. It specified a collection of gadgets, unknown
to the most privileged stratum of English society at the end of the
eighteenth century, that any clerk could take for granted a hundred
years later:

What Did George the Third Know?

He never saw a match.
He never rode a bicycle.
He never saw an oil stove.
He never saw an ironclad.
He never saw a steamboat.
He never saw a gas engine.
He never saw a type-writer.
He never saw a phonograph.
He never saw a steel plough.
He never took laughing gas.
He never rode on a tram car.
He never saw a fountain pen.
He never saw a railway train.
He never knew of Evolution.
He never saw a postage stamp.
He never saw a pneumatic tube.
He never saw an electric railway.

He never saw a reaping machine.
He never saw a set of artificial teeth.
He never saw a telegraph instrument.
He never heard the roar of a Krupp gun.
He never saw a threshing machine, but used a flail.
He never saw a pretty girl work a sewing machine.
He never saw a percussion cap, nor a repeating rifle.
His grandmother did his mending with a darning needle.
He never listened to Edison's mocking machine or phonograph.
When he went to a hotel he walked upstairs, for they had no lifts.
He never saw a steel pen, but did all his writing with a quill.
He never held his ear to a telephone, or talked to his wife a hundred
    miles away.
He never saw a fire engine, but when he went to a fire, he stood in line
    and passed buckets.
He never knew the pleasure and profit to be derived from reading *Sci-
    ence Siftings.*[40]

This list presented its gadgets as random but inviting pieces of a
world where traditional social and economic barriers to consumption
were turned upside down, though the deposed barriers of which so
much was made were those of the past, not the still-standing barriers
of the present. It suggested that even the most humdrum existence in
so exciting and bountiful an age was far preferable to a life of high
privilege in all previous and comparatively backward ones. It suggested, further, that life in the present age had generous rewards even
for the humble, who might profitably occupy all their lives contem-
plating them.

The germ of a consumer age lay in an intensified awareness not
only of a thing-filled world but of the ways in which these things were
used to mark social status. The artifacts of progress were objects of
remote worship for most people in the late nineteenth century, but their
desire to possess these objects and to participate in the lives they hinted
at was unmistakable. Popular magazines were filled with accounts of
one electrical invention after another, and with bright dreams of affluence and fantasies about the routines of work and play in a world
where electricity had transformed the lives of ordinary people. Lists
of electrical wonders were metaphors of leveling. "I expect to see
[electricity] used in every house in as simple a manner as gas or water,
so that it shall be within the reach of the poor as well as the wealthy,"
declared E. B. Dunn in the *North American Review.*[41] In *Cosmopolitan,* G. H. Knight had expressed sentiments that reached even farther
down the social stratum: "With each step in industrial progress not

only is the greater the number which can be warmed, fed, and clothed and the better are their life conditions, but in default of such progress a vast majority would not have lived at all."[42]

Implementing the Future

Performances by Wire

While imaginary far-sight machines like the telectroscope were hoped for and speculated about at length, far-sound machines actually existed. Perhaps more than any other communications invention, contemporaries considered the telephone the bellwether of a new age. According to *Electrical World*, its invention was "the voice crying in the wilderness, announcing the speedy coming of electric illumination, power transmission, transit, and metal working."[43] As the sole nineteenth-century instrument that transmitted voices across space at the moment of speech, the telephone was both a carrier of point-to-point messages to individuals and a medium of multiple address for public occasions of music, theatre, and politics. The most popular feature of the Paris Exposition Internationale d'Électricité of 1881 was such an arrangement, variously described as the theatrophone and the electrophone. From August to November crowds queued up three evenings a week before two rooms, each containing ten pairs of headsets, in the Palais d'Industrie. In one, listeners heard live performances of the Opéra transmitted through microphones arranged on either side of the prompter's box. In the other, they heard plays from the Théâtre Français through ten microphones placed at the front of the stage near the footlights.[44] Not only were the voices of the actors, actresses, and singers heard in this manner, but also the instruments of the orchestra, the applause and laughter of the audience—"and alas! the voice of the prompter too."[45]

Efforts to reach extended audiences by telephone required elaborate logistical preparations. Its application to entertainment, therefore, remained experimental and occasional. In Europe entertainment uses of the telephone were often an aristocratic prerogative. The president of the French Republic was so pleased with the theatrophone exhibit at the Paris Exposition that he inaugurated a series of telephonic soirées with theatrophonic connections from the Élysée Palace to the Opéra, the Théâtre Français, and the Odéon Theatre.[46] The King and Queen of Portugal, in mourning for the Princess of Saxony in 1884 and unable to attend the premier of a new Lisbon opera, were provided with a special transmission to the palace through six microphones mounted at the front of the opera stage.[47] The same year the manager of a theatre in Munich installed a telephone line to his villa at Tutzingen on the Starnberger Sea in order to monitor every performance and to hear for himself how enthusiastically the audience applauded. The office of the Berlin Philharmonic Society was similarly connected to its own distant opera house.[48] In Brussels, the Minister of Railways, Posts and Telegraphs and other high public officials listened to live opera thirty miles away at Antwerp.[49] Beginning in 1890, individual subscribers to the Theatrophone Company of Paris were offered special hookups to five Paris theatres for live performances. The annual subscription fee was a steep 180 francs, and 15 francs more was charged to subscribers on each occasion of use.[50] In London in 1891, the Universal Telephone Company placed fifty telephones in the Royal Italian Opera House in Covent Garden, and another fifty in the Theatre Royal, Drury Lane. All transmitted exclusively to the estate of Sir Augustus Harris at St. John's Wood, with an extension to his stables.[51] By 1896 the affluent could secure private connections to a variety of London entertainments for an inclusive annual rent of ten pounds sterling in addition to an installation fee of five pounds.[52] The queen was one of these clients. In addition to having special lines from her sitting room to the Foreign Office, the Home Office, the Board of Green Cloth, and Marlborough House, Her Majesty enjoyed direct connections to her favorite entertainments.[53]

Commercial interest in a larger, less exclusive audience was not far behind. "Nickel-in-the-slot" versions of the hookups provided by the Theatrophone Company of Paris to its individual subscribers were offered as a public novelty at some resorts. A franc bought five minutes of listening time; fifty centimes bought half as much. Between acts and whenever all curtains were down, the company piped out piano solos from its offices. In England in 1889 a novel experiment permitted "numbers of people" at Hastings to hear *The Yeoman of the Guard* nightly.[54] Two years later theatrophones were installed at the elegant Savoy Hotel in London, on the Paris coin-in-the-slot principle.[55] For the International Electrical Exhibition of 1892, musical performances were transmitted from London to the Crystal Palace, and long-distance to Liverpool and Manchester.[56] In the hotels and public places of London, it was said, anyone might listen to five minutes of theatre or music for the equivalent of five or ten cents. One of these places was the Earl's Court Exhibition, where for a few pence "scraps of play, music-hall ditty, or opera could be heard fairly well by the curious."[57]

Though it was conventionally held on both sides of the Atlantic that the telephone was used very little for amusement in the United States, telephone occasions were reported at least as frequently in America as in Europe. Even in the earliest stages of telephone development, entertainment by wire was pursued with enthusiasm in the United States. A Boston journalist recalled sitting in the Hyde Park telegraph office when the telephone was still an infant, listening to the tune "Home, Sweet Home" conveyed through a telephonic apparatus, which seemed to him a miracle.[58] By the late 1880s such listening was no longer unusual. A characteristic account reported that a virtuoso violinist from the Meriden, Connecticut, Friendly Club had played "The Cackling Hen" over a long-distance connection to Philadelphia in 1888. Applause was received from six or seven places along the line.[59] The high point of a meeting of the Editorial Association in Boston in 1890 was a telephonic demonstration in the parlor of the Boston Press Club. Delegates listened by telephone as a cornet and piano performance of "Little Annie Rooney" and other popular airs were transmitted long-distance from New York. "Later the operators made connection to the Broadway Theater and the Casino, New York, and Keith's Gaiety and Bijou, Boston; and snatches of popular opera were heard as distinctly as the cornet and piano music had been."[60] In a special experiment at the Franklin Institute in Philadelphia in 1887, W. J. Hammer recorded a live concert performance in Philadelphia into a phonograph and transmitted that recording by long-distance lines to a loudspeaking telephone in New York. There a second phonograph recorded the recording. To the delight of the Franklin Institute audience, the second recording was transmitted back to Philadelphia, where it had all started.[61]

Riding down Ocean Avenue in New York one day in 1887, a *Sun* correspondent observed two ladies waltzing in the parlor of Western Union general manager Thomas T. Eckert. One held a telephone receiver close to her own and her partner's ears, and both frequently reversed steps to avoid being tangled in the wire. The other end of the line was directly above the orchestra in the parlor of the West End Hotel, which played dance music each noon.[62] In 1890 telephone subscribers in Rochester, Buffalo, and several nearby cities heard a concert of instrumental music transmitted from New York City, along with vocal selections transmitted from Troy and Poughkeepsie.[63] Local telephone concerts were reported in Mobile, Alabama, in 1896 and Wichita, Kansas, in 1902.[64]

Informal entertainments were sometimes spontaneously organized by telephone operators during the wee hours of the night, when customer calls were few and far between. On a circuit of several stations, operators might sit and exchange amusing stories. One night in 1891 operators at Worcester, Fall River, Boston, Springfield, Providence, and New York organized their own concert. The *Boston Evening Record* reported:

> The operator in Providence plays the banjo, the Worcester operator the harmonica, and gently the others sing. Some tune will be started by the players and the others will sing. To appreciate the effect, one must have a transmitter close to his ear. The music will sound as clear as though it were in the same room.[65]

A thousand people were said to have listened to a formal recital presented through the facilities of the Home Telephone Company in Painesville, Ohio, in 1905.[66] And, portent of the future, in 1912 the New York Magnaphone and Music Company installed motor-driven phonographs that sent recorded music to local subscribers over a hundred transmitters.[67]

In 1889 a Chicago Telephone Company experiment in transmitting *The Charlatan,* a comic opera playing at the Columbia Theater, was so successful that its general manager announced plans to furnish subscribers with musical events, comedy, drama, vaudeville, and sermons by prominent preachers.[68] Manager Angus S. Hibbard pointed to the precedent set by the Wisconsin Telephone Company of Milwaukee, which had provided orchestral music from the Palm Garden resort free "as a compliment to the company's subscribers" every evening and Sunday afternoon for three years. A report six weeks into the Palm Garden experiment described it as "a distinctively twentieth-century idea."[69] In Oshkosh, Appleton, Sheboygan, Racine, and other cities where patrons gathered at hotels and clubs to hear the Palm Garden transmissions, local telephone exchanges offered continuous performances "by interspersing between the Palm Garden numbers solos on the violin, banjo, and cornet, and vocal solos, as well as a phonographic hodge-podge" from their own offices. Wisconsin general manager J. D. McLeod described his company as American "pioneers:" "Judging from the nightly demand upon us, and its popularity, there is a merchantable quantity in such entertainments; and amusement people may ultimately take it up on these lines. But as far as the Wisconsin Telephone company is concerned, it is enough to have broadened its usefulness."[70]

Telephone entertainments were not limited to musical transmissions. In March 1912 alumni at the annual Chicago Yale Club banquet heard Yale's president address after-dinner remarks to them by long-distance from a comfortable seat in New Haven. Seven hundred newspapermen gathered at the Waldorf Astoria a few months later for a joint meeting of the Associated Press and the American Newspaper Publishers' Association were treated to a special after-dinner telephone program. As each guest listened on a special receiver fitted into a watchcase, President Taft spoke from Boston, Canadian premier Robert Borden spoke from Hot Springs, Virginia, a Kipling poem was recited from Daly's Theatre, and a vocalist performed a "Southern song" from another New York theatre, the Winter Garden.[71]

Sporting events provided occasions for telephone transmission, and had inspired imaginative experiments with the telegraph. In 1884 three Nashville telegraph operators, J. U. Rust, E. W. Morgan, and A. H. Stewart, organized a "vivid view of the exact situations and plays in a game of baseball played in Chattanooga" for an audience in a Nashville hall.[72] From the playing field, one operator telegraphed each play of the game over a leased line to Nashville, where another operator announced it to the audience. The third operator moved cards bearing the players' names around a ball field painted on poster board that was visible to the entire audience.

In 1886 Detroit, Morgan & Co., as this entrepreneurial team called itself, relayed a Detroit-Chicago game to a Detroit Opera House crowd of more than six hundred by using "a huge landscape—it would have done well as a curtain—having a well painted perspective view of a baseball diamond and outfield." Slots at each playing position held the changing names of the players during the course of the game. The *Detroit Free Press* described the reaction of the audience:

> The audience during the first four or five innings of yesterday's game was wrought up to a very high pitch of enthusiasm. For instance when the operator read—with Dalrymple's name appearing as batsman—"foul fly to left," the audience fairly held its breath, and when the next instant the operator called out, "and out to White," there came a storm of applause, just such as is heard on a veritable ball field. And so it was all through the calling of strikes, balls, long hits and short ones, outs, errors and "safes," the excitement was intense.[73]

By 1889 the idea had caught on elsewhere. As a gesture of good will to the twenty-four hundred telephone subscribers of the Cleveland Company, its operators were "always informed regarding the base ball score and always ready to answer questions regarding it. They keep up with the games from inning to inning, and most of them being interested themselves are agreeable in answering all demands on their information so far. Such is the policy of the exchange."[74] Of a summer afternoon on Park Row—the newspaper capital of New York and a center for telegraphic dispatches—a reporter for the *Electrical Review* observed that baseball games in progress elsewhere were replicated by moving wooden men on specially marked blackboards "simultaneously with their counterparts of brawn and muscle on the green diamond. Here the game goes merrily on whether the original be at Staten Island, Boston, Philadelphia, or in the West, electricity again annihilating space."[75]

The 1894 games of the New York Baseball Club were transmitted by telegraph from the Polo Grounds, where they were played, to the Standard Theatre several miles away. Like the Nashville scheme of a decade earlier, simultaneous re-creations at the Standard Theatre featured a baseball diamond painted on a curtain with special holes at key locations. At the appropriate moment, figures signifying plays, calls, and players were thrust through these holes or retracted by small motors behind the curtain. When a fly was struck, for example, a large image of that insect appeared on the curtain. A coordinated system of markers and electric bells conveyed the unfolding progress of the game. "The anxiety of those present was just as great as though they had been occupying the real grand stand," commented *Electrical World,* which was amused by the excitement of the audience after each hit.[76] Less formally, tapping the telegraph wires for prized bits of sporting news was said to be a favorite pastime of New York telegraph operators.[77]

The introduction of wireless telegraphy likewise inspired schemes for relaying sporting intelligence. One of the first commissions undertaken by Guglielmo Marconi's new Wireless Telegraph and Signal Company was an assignment to report the Kingstown Regatta for the *Dublin Daily Express* in 1898. The *Express* paid the costs of reporting the race, furnished a great deal of favorable publicity to the Marconi enterprise, and alerted the rest of the world's newspapers to an important journalistic innovation.[78] The following year Marconi personally supervised the reporting of the America's (Queen's) Cup Races by wireless from the steamship Ponce in Hudson Bay for the *New York Herald* and *Chicago Times.*[79] Reports were received in New York in seventy-five seconds. Wireless reports were sent to Chicago in three to seven minutes. As each report came into the newspaper office, bulletins were posted on the front of the building for the crowds gathered outside.[80]

Church services were also an occasion for telephone transmission. From about 1894, telephone wires connected subscribers with local pulpits in towns as large as Pittsburgh and Philadelphia, and as small as Paris, Texas.[81] Inclement weather prompted the Reverend D. L. Coale to connect a large megaphone to a telephone receiver in the Anson, Texas, church auditorium where he was conducting a revival in 1912, so that those absent from services might receive the benefit of sermons and singing. More than five hundred were said to have listened to revival services, and a number of conversions were made by wire.[82]

Telephone pulpits seemed to have come earlier to British churches. An account of the inauguration in 1890 of a service in Christ Church in Birmingham with connection to subscribers in London, Manchester, Derby, Coventry, Kidderminster, and Hanley went as follows:

> When the morning service commenced there was what appeared to be an unseemly clamor to hear the services. The opening prayer was interrupted by cries of "Hello, there!" "Are you there?" "Put me onto Christ Church." "No, I don't want the church," etc. But presently quiet obtained and by the time the Psalms were reached we got almost unbroken connection and could follow the course of the services. We could hear little of the prayers—probably from the fact that the officiating minister was not within voice-reach of the transmitter. The organ had a faint, far-away sound, but the singing and the sermon were a distinct success.[83]

Subscribers in Glasgow listened to their first telephonic church service in 1892. By 1895 connections for subscribers and hospital patients had been made to the leading churches of London, including St. Margaret's, Westminster; St. Anne's, Soho; and St. Martin's-in-the-Fields and St. Michael's, Chester Square, by Electrophone Limited.[84]

An occasion that shared many features of the religious revival was the political campaign. Both events were marked by intense community discussion about the proper kind of society for people to live in, and both were strongly oral in character. Both fell to the blandishments of electricity with no resistance. In 1896, an election year that saw a prolific use of electric media in connection with political news and entertainment, the South Bend Telephone Company of Indiana connected its patrons free of charge one August evening to a tent wherein the Honorable John L. Griffiths of Indianapolis was making a Republican campaign speech to assembled supporters. "Auditors . . . at ease in their own homes" listened from one hundred residential telephones to the address, to band music, and to "the rustle and bustle incident to a large gathering of men and women. . . . congratulatory remarks, witty sallies, sarcastic jibes at the enemy, laughter, coughs." At each house the telephone was passed from ear to ear until "the sounds ceased altogether, and the receivers were hung up."[85] The occasion had been conceived as a two-way affair, with contributions from the crowd as much a part of the event as the speech of the candidate. Several months later, the parade crowd celebrating the twenty-fifth anniversary of the great Chicago fire was the featured communicator over a long-distance hookup to the whole country:

> The telephone building in Washington Street arranged a large trans-mitter over the street, so as to catch the cheers and music of the proces-sion as it passed. The transmitter was connected to all the telephone lines in the country, and as the procession passed its cheers were heard throughout the Union.[86]

The telephone could reproduce the heinous as well as the holy, and some communities were interested in both. Public justice pursued in the face-to-face environs of the courtroom was occasionally ex-tended beyond its walls. The sensational murder trial of Reginald Bir-chall in Woodstock, Ontario, in 1890 attracted people from every neighboring town to a courtroom scarcely large enough to accom-modate the official participants in the trial, alongside journalists from London, the United States, and all Canada. An enterprising local tav-ern keeper arranged to install a transmitter above the judge's bench in the courtroom. This was connected to twenty receivers in his tavern, each of which he rented for twenty-five cents an hour to the overflow of the curious, and presumably the thirsty. Four tubes were connected to a private room for ladies. All were kept busy.[87]

Systems of Electric News

Early in 1889 the *Electrical Review* summarized "In the Twentyninth Century: The Day of an American Journalist in 2889," a Jules Verne short story that portrayed a great American editor one thousand years thence:

> The editor rules the world; he receives ministers of other governments and settles international quarrels; he is the patron of all the arts and sciences; he maintains all the great novelists; he has not only a telephone line to Paris but a telephote line as well, whereby he can at any time from his study in New York, see a Parisian with whom he converses.
>   Advertisements are flashed on the clouds; reporters describe events

orally to millions of subscribers; and if a subscriber becomes weary, or is busy, he attaches his phonograph to his telephone, and hears the news at his leisure. If a fire is raging in Chicago, subscribers in New York may not only listen to the description of an eyewitness, but by the telephone may see the wire.[88]

Most predictions of the future of newsgathering were not so dramatically detailed, but the theme of control at a distance so dear to the hearts of scientists and engineers was here extended to the world's understanding of itself through a process of centralized electrical monitoring. The Faustian impulse to embrace the whole world in the nerve net of electricity already had created telegraphic wire services for this purpose. Writers elaborated this original model by speculating on the talents of the newest electric media. "It becomes imaginable that one operator on a typewriter keyboard may be able to set type or work linotype machines simultaneously in a dozen cities at any distance," a popular journal forecasted in 1894.[89]

Innovative systems were often recapitulations. Telegraphic news reached nineteenth-century publics not only through the traditional medium of the printed newspaper, but also in bulletins hastily scrawled from the latest dispatches and posted outside newspaper offices. A bulletin board electrically automated for this purpose was exhibited in New York in 1888. It consisted of a row of horizontal windows through which messages were spelled out by a series of separately revolving wheels inscribed with alphanumeric characters. The electric bulletin board was "not intended to have a record of the news it conveys, but is designed merely to satisfy the eagerness for news."[90]

The *New York Times* led its sister newspapers in the regular use of wireless telegraphy. For European stories in its Sunday edition, for which mails were too slow and undersea cables too costly, the *Times* had come to rely on wireless by 1908. Four years later it was receiving "practically all of its daily foreign news service by wireless telegraphy," a then remarkable stream of about twenty thousand words a week.[91] This exchange between the old world and the new required a cooperative network of cables, telephones, and wireless stations, and was said to have broken a number of speed records in overseas wireless transmission.

The distribution of presidential election returns in the late nineteenth century was the most ambitiously organized American effort to use new electric technologies to deliver the news. Election returns had been distributed by the telegraph since its invention, but the telephone added speed, immediacy, and convenience. Early telephone distribu-

tion depended on a backbone of telegraphic returns and used supporting visual technologies such as the stereopticon, the kinetoscope, and
the electric searchlight. From 1892, a growing demand for quick and
comprehensive election statistics was met and doubtless augmented by
unifying old and new networks for increased capacity and flexibility.

The first coordinated system of telephone returns was organized
for the presidential election of 1892. During the previous election,
American Bell Telephone Company president Howard Stockton had
invited guests to his Boston home to hear the returns come in over a
special telephone wire on election night.[92] For the 1892 election, the
telephone companies of New York and Chicago arranged to forward
returns coming in to them from telephone and telegraph wires across
the country to all interested New York and Chicago clubs and hotels.
Information was sytematically exchanged between these two major cities over newly laid circuits through Milwaukee. Some telephone bulletins were received as much as ninety minutes in advance of bulletins
from local telegraph offices.[93]

Offices at 18 Cortlandt Street in New York were the central distributing points for returns in a system engineered by General Superintendent Angus S. Hibbard. More than 380 bulletins were sent outward to a territory including New England, most of the Middle Atlantic
States, Wilmington, Washington, and Baltimore on the south, and Chicago and Wisconsin to the northwest.

> A long distance operator was stationed at every place where returns
> were furnished, and as fast as they were received they were written
> down in crayon on white sheets of paper about 12 by 18 inches in size,
> at the top of which was printed "Long Distance Telephone Bulletin."
> The telephone company also furnished its patrons with printed cards
> showing the vote of previous years and provided with blank spaces for
> this year's returns.[94]

Bulletins were sent in to Cortlandt Street from newspapers, police
headquarters, and other official sources. In some cases long-distance
operators were stationed at individual voting precincts. At Cortlandt
Street, doors were stripped off offices so messengers could move rapidly back and forth, and incandescent lamps were positioned over every
instrument and switchboard to give operators plenty of light. Editors
at centrally located desks condensed incoming information for out-
going distribution, and carefully edited the "considerable partisan feeling" that colored many incoming reports.

Not until the presidential election of 1896 were long-distance lines plentiful enough for telephone companies to organize a national network for gathering and distributing election returns. Here, too, extensive telegraphic support was still necessary. At American Telephone and Telegraph's New York headquarters on election night in 1896, more than a hundred persons received, edited, and forwarded bulletins to local subscribers. Between half-past six o'clock in the evening and one o'clock in the morning, a second staff of operators accepted more than twelve hundred long-distance bulletins from designated points. These were culled, edited, compared, and relayed to New Yorkers faster than competing telegraphic services were able to keep even the press informed.[95] Three different return services were organized for election night. Operators in the first telephoned returns to local hotels and clubs, where they were copied onto large sheets and displayed by stereopticon or magic lantern to waiting crowds. A second service telephoned the latest bulletins to 350 local exchanges. Groups of twenty or so subscriber lines were connected to special telephones at these exchanges, over which operators read the bulletins as quickly as they arrived. In the third service, fifty operators handled more than fifteen hundred incoming calls of inquiry.[96]

For the election of 1900, AT&T collected local returns from telephone companies across the country, compiled summary bulletins, and circulated updated returns back through the same network of local exchanges. In New York City, the New York Telephone Company provided services to thirty-two Manhattan clubs and hotels and thirty-five country clubs, hotels, and associations in Westchester County, and connected individual subscribers to a hundred special stations where operators with the latest information answered their inquiries. Stereopticon bulletins were also displayed at the principal exchanges. In New York, less demand for the return-reporting services of Western Union than in previous elections was attributed to the improved quality of telephone equipment and services.[97]

With the cooperation of AT&T, the telegraph companies, and their local and state networks, the Chicago Telephone Company relayed returns within a minute of receipt to an estimated twenty-five hundred subscribers in hundreds of clubs and houses where private telephone parties were in progress. A General Electric searchlight projected by the *Record* from the top of the Chicago Masonic Temple did not fulfill the fondest hopes of sky-writing enthusiasts, but it did communicate a message:

It had been announced that McKinley's election would be signaled by a steady horizontal sweep of light from left to right and right to left, while if Bryan were successful the beam was to be swung in a circle around the horizon, with a vertical up-and-down motion. As long as the result was in doubt the agreed signal was a steady, vertical ray. Many watchers throughout Cook County thus received their first news of the result.[98]

In 1912 the distribution of election returns was still provided free of charge and without interruption to regular telephone service, but it had been much refined and perfected. At the Chicago Telephone Company's downtown office, incoming national returns from telegraph and telephone hookups and local and state returns provided exclusively by telephone were forwarded to special editors who prepared summaries that were passed orally to subscribers by readers at each local exchange. Fourteen local wires reached twenty-eight city exchanges. Twenty suburban and long-distance circuits were connected to fifty-five northern Illinois towns and some as far south as Danville, Kankakee, and Rockford. The public was notified in the newspapers to call "Election News" for the latest bulletins, each approximately one page long and requiring two minutes to read. At the end of each bulletin readers announced, "Please hang up your receivers, another bulletin will be read in ten minutes."[99] Bulletins were read continuously to one group of subscribers, then another, until new bulletins came in. Special equipment prevented interference or interruption from subscribers attempting to speak or to signal the operator.

What did it mean, this novel network of the latest election intelligence, to its beneficiaries? It did not seem to them to be simply an extension of the telegraph. Dismissing the familiar wire service network as "very largely a mechanical operation, which a child can comprehend," the *Manchester* (New Hampshire) *Union* concluded that "the collection of the returns from Kickapoo, Arizona; Masardis, Maine; Laredo, Texas, and their dissemination among millions of people before they retire to the privacy of their homes is another matter."[100] Impressed by estimates that perhaps half a million persons had received telephone returns in their homes and offices, the *Union* marveled:

The news was, literally speaking, scooped up in this great telephone net and talked into one's ears from unexpected distances. . . . Thousands sat with their ear glued to the receiver the whole night long, hypnotized by the possibilities unfolded to them for the first time. . . . If we can hear hundreds of thousands of people scattered over this broad land speak, why can we not in time produce other wonderful results now deemed impossible?[101]

Telephonic news seemed poised to overtake telegraph news, an American electrical journal commented in 1895:

> Already in large cities, the ordinary subscriber uses his telephone ten times a day. . . . Moreover, we record this week the use of the long distance telephone wholesale for [political] convention news purposes, thus sapping the vitals of the telegraphic news systems of the Associated and United Press.[102]

Though the picture was not quite that grim, since the telephone and telegraph would learn useful ways of cooperating in the offices of the wire services, newspapers were indeed enamored of the excitement of telephone news. The capacity to communicate the thunder of events directly to an audience with an immediacy greater than that of the telegraph or newspaper alone distinguished the new electric media from the old, even as the new media were pressed into the service of the old. In spite of the American Bell Company's policy against permitting private individuals or companies to send news by telephone wire, late-nineteenth-century newspapers routinely featured late-arriving telephonic dispatches with bold headings announcing precisely this means of transmission.[103]

By reproducing and simultaneously transforming the telegraphic information network, the telephone distribution of breaking news was part of the transition from a passing world. The crowds that gathered in the streets to celebrate McKinley's victory in 1896 had little inkling in that euphoric moment that their descendants would learn the results of great political contests in sedately familiar living rooms. *Harper's Weekly* described the world that was passing, and, without recognizing them, some of the instruments of its transformation:

> Such a crowd as tramped and cheered and roared up and down Broadway election night, and surged about every building where a calcium-light was throwing election returns upon a screen, has never before been seen in New York. . . . The crowds on upper Broadway were entertained as well as instructed; between bulletins on one screen there was an exhibition of the vitascope, and as the scenes were flashed upon it the shouts of laughter and merriment rose above the din of horns and rattles.[104]

Other efforts to distribute the news were organized from time to time. In the farming country of the Midwest at the turn of the century, weather reports were regularly read over the wires.[105] In April 1898, when all signs pointed to the entry of the United States into Cuba's war against Spain, the general manager of the Chicago Telephone Company promised that every one of the company's fourteen thousand subscribers would be notified by operators within twenty minutes of any official declaration of war. "Our idea in doing this is to inform our patrons of the declaration probably quicker than they would otherwise get the news," Angus S. Hibbard explained in an official announcement: "We are a quasi-public corporation and we rather consider it our duty to act this way. . . . Of course we will try to guard against any canard, but in no event will we assume responsibility for the news as we send it out. Our operators will simply tell the subscribers that we have received it as news."[106]

Ways of using the telephone to get the latest news were also improvised without any professional assistance at all. On party line systems one found "listeners all along the wire for every scrap of conversation going. So a whole countryside may learn that the doctor is on his way to Mrs. Brown, Mrs. Jones or Mrs. Robinson."[107] Not all of these listeners were interested only in local news. In 1905 *Telephony* reported that every afternoon in Evanston, Illinois, a subscriber called a prominent business house to inquire, " 'Well, what's the news today? Somebody just said that the Atlantic won the yacht race, is that so? Has anybody resigned from the cabinet today? How did the Chicago-Pittsburgh game come out? Anybody hurt in the trolley collision? Do you know what day the Cunard liners sail for Europe?' "[108]

Telephone Diffusion: A Proto–Broadcasting System

In the late nineteenth century, single events such as a declaration of war, a baseball game, a church service, or a concert were transmitted by new technologies with unprecedented immediacy to scattered audiences *on occasion*. Although modern media transmit content of a similar kind, late-nineteenth-century telephone occasions otherwise bear little resemblance to twentieth-century mass media programming. Nineteenth-century telephone occasions were derived transmissions of independently occurring events and were intended to extend the primary audiences of the pulpit, stage, concert hall, and playing field. Wholly invented programming, by contrast, is a distinctive social feature of electronic mass media.

Commercial efforts to enlarge audiences electrically for some regularly repeated occasions in the late nineteenth century were generally of short duration; the audiences they attracted were small. Electrophone parties in Britain were said to be a pastime of the idle rich, not the humble poor.[109] Electrophone Ltd., one of the sturdier British companies to take up regular telephone transmission, piped sermons from the most prestigious pulpits and plays from the most prestigious theatres to London's leading hospitals for the edification of affluent patients, and to occasional private residences as well. Nevertheless, twelve years after its incorporation, Electrophone had a regular subscriber audience of barely six hundred.[110]

But from 1893 until after World War I, when a number of private companies and national states began to create radio broadcasting systems, an organization in Budapest was a remarkable exception to the usual pattern. This was the Telefon Hirmondó, which for almost a generation transmitted daily programming over telephone wires to supplement the regular telephone service of more than six thousand subscribers. *Hirmondó* was a Magyar term for the crier who shouted the news from the center of the medieval village for all to hear. Today it denotes a radio announcer. Its semantic transformation followed a path directly through the career of the Telefon Hirmondó. For twenty years the Hirmondó's audience received a full daily schedule of political, economic, and sporting news, lectures, plays, concerts, and recitations. The language of the Telefon Hirmondó was Magyar, the language of Hungarian nationalism. In operation, the Telefon Hirmondó was a closed and exclusive system of cultural communication among the Hungarian elite during the last decades of Magyar power before World War I, a fact that appears to account for both its economic and its cultural staying power.

The Telefon Hirmondó was the brainchild of Tivadar Puskás, a Hungarian engineer who had worked on Thomas Edison's staff of inventors and researchers at Menlo Park. To Puskás, according to Edison, belonged the original credit for suggesting the concept of the telephone switchboard that made the telephone a powerful and practical means of communication. Accounts of the Telefon Hirmondó were followed with interest in the British and American press, and a short-lived imitation of it appeared in the United States. It provided perhaps the only example of sustained and systematic programming in the nineteenth century that truly prefigures twentieth-century broadcasting systems.

The origins of the Telefon Hirmondó lay in the novel and popular theatrophone exhibition that Puskás helped mount at the Paris Exposition Internationale d'Électricité in 1881. The following year he staged his own theatrophone demonstration in Budapest by transmitting a National Theatre opera performance to a nearby grand ball.[111] In the meantime, Puskás's brother, Ferenc, acquired the first telephone concession in Budapest, and the Puskás family hired Nikola Tesla, a longtime friend, to engineer its construction.[112] The Budapest telephone system prospered under Ferenc Puskás, and in 1892 Tivadar Puskás, who had played a minor role in some of the more exciting electrical developments of the age and knew many of its foremost inventors and engineers personally, returned to Budapest to implement his own remarkable idea of a Telefon Hirmondó. The first program was transmitted from the central telephone exchange to one thousand regular telephone subscribers in 1893. Within weeks of the inception of the Telefon Hirmondó, Tivadar Puskás was dead. His creation outlived him by almost a quarter of a century.

At first the Telefon Hirmondó's programming consisted of news summaries read at the beginning of each hour and immediately repeated. Silence reigned until the next hour's transmission. Five months into the new experiment, *Science Siftings* reported:

> The news collector does his work in the night, and having his budget filled he takes his place in the central office at nine in the morning and begins to tell his story, which is given in a telegraphic style, clear, condensed, and precise. In five minutes after the first delivery the budget of news is repeated, in case some of the subscribers may not have heard. It consists for the most part of home events and news of Hungary. At ten o'clock the foreign news is given, and after eleven the doings of the Hungarian Parliament. Various items of city news are given during the day.[113]

News in the daytime was balanced by cultural programs in the evening—perhaps a report of a lecture at the Hungarian Academy, or the recitation "with all due emphasis" of a new poem.

Efforts to transmit music met with poor success and provided the first indications of a problem that increased with the listening audience. Simply stated, the addition of subscriber outlets diminished the volume of sound for every subscriber. When control of the Hirmondó passed out of the hands of the Puskás family in 1894, a new distribution system that bypassed the regular telephone network eliminated this and other technical problems.[114] The new company was granted the same right to place its wires as the telephone and telegraph companies. By 1900 the Telefon Hirmondó employed over 150 people in its offices at 22 Megrendelhető Rákóczi, on one of the "finest avenues" in Budapest.[115]

The news operation was like that of any newspaper. News from abroad came by telegraph. Local news was assigned to a staff of twelve reporters. A special staff assigned to the galleries of the Hungarian and Austrian Houses of Parliament forwarded half-hourly reports of the latest developments.[116] Galley proofs of every story were printed by hand roller presses in parallel columns on sheets of paper two feet by six inches. Several sheets constituted the daily program. The work of the "stentors" who read the news was thought to be so exhausting that they were rotated at ten-minute intervals in groups of four.[117]

By 1896 the daily program of the Telefon Hirmondó had achieved virtually its final form. This version is translated from a German publication, which published it in full:

"Telefon Hirmondó" Order of the Day

9.30–10.00 Daily calendar, Vienna news (telephone report), latest telegrams (arrived during the night), train departures listed in the railway gazette

10.00–10.30 Report of the stock exchange

10.30–11.00 Review and summary of the day's newspapers, telegrams

11.00–11.15 Report of the stock exchange

11.15–11.30 Theatre news, sport and local news

11.30–11.45 Report of the stock exchange

11.45–12.00 Parliamentary, foreign and provincial news

12.00–12.30 Parliamentary, military, political and court news

12.30– 1.30 Report of the stock exchange

1.30– 2.00 Repeat of the most interesting news read so far

2.00– 2.30 Parliamentary and municipal news, telegrams

2.30– 3.00 Parliamentary, telegraphic and local news

3.00– 3.30 Report of the stock exchange

3.30– 4.00 Parliamentary news, exact zone time, weather report, medley

4.00– 4.30 Report of the stock exchange

4.30– 5.00 Vienna news (telephone report), political economy

5.00– 5.30 Report on theatre, art, literature, sport and fashion, theatre and amusement notices, calendar for the next day

5.30– 6.00 Legal, local and telegraphic news

6.00– 6.30 Repeat of the most interesting news read so far

about 6.00 Presentation of the Royal Hungarian Opera House, or performance of the Folk Theatre

If nothing is heard at this time, this is because of a scheduled:

7.00– 8.15 Pause

8.15– 8.25 Report of the stock exchange

8.25– 9.00 Concert of the *Telefon Hirmondó*

9.00–10.00 Latest telegraphic, local and market report

10.00–10.30 The above news will be presented here at the conclusion
of the performance of the Folk Theatre

Thursday evening

6.00– 6.45 Children's concert

Program for Sundays and holidays

11.00–11.15 Daily calendar, report of the stock exchange

11.15–12.00 Review and summary of the day's newspapers, telegrams, gazette

12.00–12.30 Municipal news, sport and theatre news

12.30– 1.00 Local and Vienna news (telephone report)

4.00   Grand concert of the *Telefon Hirmondó*[118]

Beginning about 1896, nationally known authors read serial install-
ments of their novels, to the delight, it was said, of the female audience. A popular innovation the following year was a special time signal, a powerful oscillator that buzzed for precisely fifteen seconds before
each hour.[119]

Photographs and illustrated advertising posters show that subscribers listened to the Hirmondó through two small round earpieces hanging from a diamond-shaped board mounted on the wall.[120] The audience for which the service was intended apparently possessed wealth,
education, and leisure. Its cultural relaxations were those of the opera
and the theatre. Its attachment to sport was aristocratic. The latest intelligence from the principal Hungarian and Austrian racetracks, the
cycling and automobile track, and the rugby field and billiard table
was "flashed over the wires the moment the results are known."[121] Its
children received proper cultural exposure in a weekly children's program of short stories, songs, recitations, and instrumental music.[122]

The Hirmondó devoted the largest share of its programming to the
conditions and exigencies of the financial world. Even on Sundays and
during evening programs with an artistic and performing emphasis, due
attention was given to the stock exchange. Subscribers were kept posted
about developments in the Hungarian and Austrian exchanges and the
foreign exchanges, including Wall Street and London.[123] News was
also communicated directly from the agricultural districts of the country for speculators in corn and wheat.

Our hypothesis that the audience of the Telefon Hirmondó was composed of elite and influential Budapest citizens is confirmed for the few subscribers whose names we have—the prime minister and all the members of the Hungarian cabinet, the mayor of Budapest, and Moric Jokai, a Magyar author and celebrity whose work was often featured by the Hirmondó.[124] A partial street-wiring diagram published in an 1897 Hungarian encyclopedia shows Telefon Hirmondó connections to what was then and is still a wealthy section of the city, an area of elegant avenues, fine hotels, government buildings, and luxurious private residences within a famous half-circle of boulevards on the Pest bank of the Danube River.[125] The Telefon Hirmondó was also connected with doctors' waiting rooms, large coffeehouses and cafés, hospitals and hotels, merchants' and lawyers' offices, barbershops and dentists' parlors.[126] By 1896 the Hirmondó boasted six thousand subscribers, but this figure represented barely one percent of the population of Budapest.[127] The number of subscribers remained almost constant until 1917, after which reports of the Telefon Hirmondó dropped out of the foreign press. The audience of the Hirmondó was probably larger, since each household may have represented several listeners, and semipublic installations seemed to attract many different listeners. A traveler's account from 1908 explained how this worked:

> "You may be seated as I was in the reading-room of one of the hotels or in a large coffee-house, when suddenly a rush is made for a telephone-looking instrument [the Telefon Hirmondó] which hangs from the wall. In time perhaps you will become one of these 'rushers.'"[128]

Nevertheless, subscription figures were small, a fact that cannot be accounted for by price, since the installation of the Hirmondó apparatus was free, and subscriber fees were only a penny a day. Not even the fact that the Hirmondó transmitted exclusively in Magyar, a minority language, explains the size of its subscribing audience, since Magyar was also the official language of Parliament, the universities, and the high courts. (Foreign-language lessons were regularly featured in Hirmondó programming, but not instruction in Croatian or Slovak, the languages of peasant peoples within Hungarian boundaries. Subscribers learned French, English, or Italian—useful languages to the cosmopolitan leisured, to merchants and diplomats.[129]) The most likely hypothesis is that Hirmondó connections were officially limited, since no citizen received regular telephone service without government permission.[130] The Hirmondó was authorized to offer its services by an exclusive government license; its programming was identified with the ruling Magyar elite.

Articles on the Telefon Hirmondó appeared frequently in European and American journals during the 1890s and early twentieth century. It was featured in the expert press, in penny weeklies with mass appeal, and in sober middle-class monthlies. Popular comment about the Hirmondó associated it with leisurely Continental lifestyles, since British and American observers often remarked with disapproval that the length of the connecting wires made it possible for subscribers to recline while listening to its programs.[131] Little is known of the Telefon Hirmondó following World War I, during which most of its exterior installations were destroyed. In 1925 the Telefon Hirmondó and Hungarian Radio Broadcasting were combined into a single organization, and the Hirmondó became merely a wire-diffusion agency for studio-broadcast programs.[132]

In the United States at least one brief experiment was directly inspired by the Hirmondó. This was the Telephone Herald of Newark, New Jersey. After sampling the Telefon Hirmondó on vacation in Budapest, a former *New York Herald* advertising manager, M. M. Gillam, set about organizing a similar enterprise in the United States.[133] Gillam and William E. Gun, builder of the battleship *Oregon,* organized the New Jersey Telephone Herald Company with promises of financial backing from a wealthy New York coal magnate. Just as the service was scheduled to begin operating, in March 1911, the New York Telephone Company reneged on its contract to lease wires to the Telephone Herald, which it regarded as a competing public utility.[134] After six months of legal wrangling, the New Jersey Public Utilities Commission held the telephone company to its original agreement. The Telephone Herald inaugurated service on October 23, 1911, with the following daily program:

Daily Program of the "Telephone Herald"

8.00	Exact astronomical time
8.00– 9.00	Weather, late telegrams, London exchange quotations, chief items of interest from the morning papers
9.00– 9.45	Special sales at the various stores; social program for the day
9.45–10.00	Local personals and small items
10.00–11.30	New York Stock Exchange quotations and market letter
11.30–12.00	New York miscellaneous items
Noon	Exact astronomical time

12.00–12.30 Latest general news; naval, military, and Congressional
             notes
12.30– 1.00 Midday New York Stock Exchange quotations
1.00– 2.00 Repetition of the half-day's most interesting news
2.00– 2.15 Foreign cable dispatches
2.15– 2.30 Trenton and Washington items
2.30– 2.45 Fashion notes and household hints
2.45– 3.15 Sporting news; theatrical news
3.15– 3.30 New York Stock Exchange closing quotations
3.30– 5.00 Music, readings, lectures
5.00– 6.00 Stories and talks for the children
8.00–10.30 Vaudeville, concert, opera[135]

The schedule of items presented by the Telephone Herald was faithfully modeled on the Telefon Hirmondó's "order of the day." The style of program presentation was also familiar:

> With his mouth between the [two] transmitters the stentor reads an item, says "change," then immediately begins upon another. As the stentors have had special courses in distinct enunciation every word can be clearly heard. The work is so exhausting that one man reads only fifteen minutes, then rests for forty-five minutes while others take his place.[136]

No programming was originated by the Herald itself, besides occasional concerts performed in a room set aside for that purpose. The Telephone Herald also had no reporters of its own. Its newsroom was entirely devoted to editorial functions:

> There is the usual barn-like room meagerly furnished, with dirty windows guiltless of shades, the floor littered with waste paper and the regulation paste pot that has not been cleaned since the year one. In these familiar surroundings a couple of editors smoke cigarets and clip the morning papers, go through press reports, proofs from a local evening paper, and correspondents' manuscripts, receive telephone messages, condensing everything to the uttermost, two hundred and fifty words being the maximum limit for the most important items.[137]

Within a month of its beginning, the Telephone Herald had acquired more than a thousand subscribers, each of whom paid a nickel a day for its services. Among them were not only individuals, but a Newark department store, whose use of the Telephone Herald as a promotional draw anticipated efforts by Gimbel's and other stores several years later to attract patrons with wireless hookups.[138] The success of the department store encouraged a local restaurant to make connection. Reportedly, its customers were so interested in the news that they ceased to find fault with their food.[139] Several clubs also subscribed.

The capital reserves of the Telephone Herald proved to be unequal to the popular demand for it, and the legal contest with the telephone company had frightened off investors. With depleted financial resources, the Herald was unable to install equipment fast enough to meet its subscription orders. After three months, twenty-five hundred subscriber contracts had been drawn up, but the number of installations was not much over a thousand.[140] Soon the financial strain began to show. Employees were irregularly paid. The musical service ended abruptly one afternoon when the musicians refused to play any longer without salary. The newsroom staff of two editors and four stentors departed a month later. Lacking capital funds, the service was suspended, and then entirely disbanded.[141]

The history of the Telefon Hirmondó and its admiring imitator, the Telephone Herald, demonstrates that the notion of transmitting regular news and entertainment programming to large audiences existed well before the advent of twentieth-century wireless broadcasting. The existence of these two precursors did not generate any popular shock of recognition, however, or nurture any expert consensus that their efforts marked an inevitable path to the future. While the public was generally confident that something fantastic and all-embracing was germinating among the many remarkable contraptions of electrical communication, the boundaries of immediate possibility appeared much narrower to those closest to the technologies involved.

The historical development of mass broadcasting ahead of cable programming, which the Hirmondó more closely resembled, might have been reversed if radio had not been invented at a time when wire diffusion was still largely experimental. It was not immediately realized how significant a departure from telephony and telegraphy radio would be, however. As late as 1921, Walter Gifford, then four years away from assuming the presidency of AT&T, had difficulty visualizing separate roles for wired and wireless media in the twentieth century. He recalled in 1944:

> Nobody knew early in 1921 where radio was really headed. Everything about broadcasting was uncertain. For my own part I expected that since it was a form of telephony, and since we were in the business

of furnishing wires for telephony, we were sure to be involved in broadcasting somehow. Our first vague idea, as broadcasting appeared, was that perhaps people would expect to be able to pick up a telephone and call some radio station, so that they could give radio talks to other people equipped to listen.[142]

Late in 1921 an internal Bell Telephone memorandum had projected the future of broadcasting simply as the transmission of important occasions, such as Armistice Day ceremonies or presidential inaugurations:

> We can imagine the President or other official speaking in Washington . . . and that his voice is then carried out over a network of wires extending to all the important centers of the country. . . . In each city and larger town there are halls equipped with loud speaking apparatus at which the people in the neighborhood are gathered.[143]

If historical events had occurred in a different order and wire diffusion had been left unchallenged to develop at its own pace, that pace might have been a slow one. Through a combination of technical and economic constraints, wire diffusion might have evolved to suit the needs and interests of privileged minorities, filtering down only gradually to a wider population. By making the delivery of content cheaper and more democratic, wireless communication made mass audiences possible for electric media, and accelerated the development of programming of all kinds.

The Telefon Hirmondó was a hybrid of newspaper practices, conventional modes of oral address, and telephone capabilities that anticipated twentieth-century radio. In operation it was a transitional form using conservative techniques that looked backward to newspaper methods for gathering information, which it presented as spoken newspaper items. In its time it was seen as a novel newspaper form, but it was radically forward-looking in its continuous and regularly scheduled programming, the origination of some programs from its own studios, and the combination of news and entertainment in the same service. No other telephone diffusion experiments embraced a system of regular, timely programming like that of the Hirmondó. Most were limited simply to the reproduction of full-length "occasions."

Epilogue

This has been a study of how groups with competing logics of experience entertained new technological possibilities in communication in the late nineteenth century. The central players in the drama were experts charged with constructing this particular technological world and publics who expected to live in it. Technological worlds are so much with us that we seldom question the creation myth in which technologists are champions of novelty, change, and progress, but never critics. Only nontechnologists are thought occasionally to doubt the wisdom of a world that technologists have put in place. This opposition between technologists and nontechnologists is a false one in many respects. Technologists are not solely members of professional groups; they are social actors with a variety of loyalties that may not always be perfectly congruent with professional goals. Even their professional roles cannot be fully understood without attention to their efforts and aspirations as members of families, citizens of countries, and possessors of gender and race. They differ from other groups partly in being more attentive to discourse about technology, and in this guise they offer a window on the way in which an entire society confronted the introduction of electricity in the late nineteenth century.

   In its most tangible aspects, electricity came to existing groups less as the transformative agent of its own mythology than as a set of concrete opportunities or threats to be weighed and figured into the pursuit of ongoing social objectives such as preserving class stability or moving upward socially. Experts and publics greeted a new world of electricity by elaborating an old one. New electrical inventions and ways of thinking about electricity were given shape and meaning by being grafted onto existing rules and expectations about the structure of social relations. Experts, and not only experts, were content to believe that the form in which the world offered itself was the best and most natural one, subject to appropriate adjustments by electrical technology but not to fundamental rearrangement. In their concern to preserve a familiar social order whose advantages to themselves were enhanced by technology, experts were bellwethers of their society, not apart from it.

Electrically transformed communication thus offers a keenly focused view of the process of social adjustment around new technology, which is an occasion for introducing new rules and procedures around unaccustomed artifacts to bring them within the matrix of social knowledge and disposition. Since communicative practices always express social patterning, any perceived shift in communication strikes the social nerve by strengthening or weakening familiar structures of association. From this perspective, the early social history of electric media must be reckoned backward from the period when the industry organized to build, manufacture, and promote radio sets for mass audiences to that period during the late nineteenth century in which the telephone entered the home and the electric light lit up the night sky. Every essential concern about the promises and risks of broadcasting as a novel communicative form had been articulated long before in discussions of earlier electric media.

The life-giving and -destroying potential of the ether made it ideally suited to carry the freight of social fantasy. The key dimensions in the world of electrical imagination were three. Electrical discourse shaped itself first to the human body, the frame in which experience is absorbed and measured by every individual according to complex cultural codes. Ways of viewing the body and its activities help mark an essential, socially constructed distinction between nature and culture. In the late nineteenth century, the intellectual establishment represented by scientific thought opposed nature, and its representative, the body, to the culture of technical knowledge as a product of literate modes. The parties to this debate were divided between the authority of the body-as-touchstone and disembodied theory in matters of knowledge about the nature of the world. Such was the cogency of bodily authority that even sophisticated observers entertained magical convictions about the relationship between electricity and their bodies, though they couched these beliefs in terms acceptable to the scientifically nurtured and declared their respect for the purely rational imagination. New technology and preferred literate habits of scientific thought failed to displace the body from its central position as an arbiter of experience. Practices in which the body was relied on to evaluate an electrified world included its adornment with electric devices of all kinds, tales of freaks with special electrical powers, and tales of sickness induced by electricity and cures effected by it. The effort to place electricity in a cultural relationship to the body also included speculation about the significance of the physical distance from direct experience that electrical communication made possible.

After the body, the second framework that structured the social meaning of electricity was the immediate community—the family, the professional group, gender, race, and class. New electric media changed the cues, or authenticating fictions, by which groups in these categories estimated the trustworthiness of those with whom they had dealings, and around which they managed strategies of deception and face-saving that also supported social stability. New media also altered real and perceived social distances between groups, making some groups more accessible and other groups less so to still other groups. If electric communication seemed to threaten certain boundaries of family, gender, and nation, its implementation was also a condition for advancing professional status and establishing a highly serviceable barrier between experts and laymen. One way to maintain a social boundary is to charge a high fee for admission. Competition between experts and laymen for the right to exercise interpretive authority over anomalous and problematic events, to cultivate an esoteric language, to elevate literate over oral competence, and to certify professionals provided the most visible instances of the negotiation of authenticity within a larger community.

The third realm that shaped nineteenth-century social discourse about electricity was the unfamiliar community. The subject of this discourse was the use of electricity to organize and regiment the world outside the family, the world outside the expert fraternity, the world outside the male fraternity, the world outside the middle class, the world outside the nation. By fantasy substitutions of a thing-filled world for painful resource and ideological conflict, homogenizing electrical technology would render politics obsolete, according to a familiar scenario. In the absence of that world in the meantime, experts with technical command of new media sometimes used them to deceive the despised and unwary, and to conduct the battle between the sexes on a new front. Like much of the rest of the culture to which they belonged, electricians entertained visions of one-way, one-world homogeneity in which difference was deviance. Electric media were central to this work of cognitive imperialism, in which Western civilization was the center of a stage play for which the rest of the world was an awestruck audience. If telephony and telegraphy were imaginatively elaborated in the telectroscope, a characteristic fantasy invention of extended cultural homogeneity, telephony and telegraphy were organized in compliance with existing social boundaries, providing avenues for mostly familiar social contact, news, and entertainment.

Such accounts provide a framework for understanding early electric communication not in the record of cumulative improvements in engineered devices, nor in institutional narratives of how emerging industries built and marketed new devices of communication, but in the history of continuous concern about how new media rearrange and imperil social relationships. That concern is as familiar as the debate about what television brings into our living rooms. Early uses of technological innovations are essentially conservative because their capacity to create social disequilibrium is intuitively recognized amidst declarations of progress and enthusiasm for the new. People often imagine that, like Michelangelo chipping away at the block of marble, new technologies will make the world more nearly what it was meant to be all along. Inevitably, both change and the contemplation of change are reciprocal events that expose old ideas to revision from contact with new ones. This is also how historical actors secure in the perception of continuity are eternally persuaded to embrace the most radical of transformations. The past really does survive in the future. Perhaps it surprises us as much as our ancestors would be surprised by what has become of the future about which they dreamed.

Notes

Note: The city of publication is given for all periodicals except those published in New York, which comprise the majority of citations (e.g., *Electrical Review, Electrical World*).

1 Inventing the Expert

1. Thomas P. Hughes, *Networks of Power: Electrification in Western Society, 1880–1930* (Baltimore: Johns Hopkins University Press, 1983); Alfred P. Chandler, *The Visible Hand: The Managerial Revolution in American Business* (Cambridge, Mass.: Belknap Press, 1977).

2. David Noble, *America by Design: Science, Technology and the Rise of Corporate Capitalism* (New York: Alfred A. Knopf, 1977).

3. For general background on status issues affecting engineers and other professionals in nineteenth-century America, see Burton J. Bledstein, *The Culture of Professionalism: The Middle Class and the Development of Higher Education in America* (New York: W. W. Norton, 1976). An account of a twentieth-century confrontation between professionalism and social responsibility is Edwin T. Layton, Jr., *The Revolt of the Engineers: Social Responsibility and the American Engineering Profession* (Cleveland: Press of Case Western Reserve University, 1971), a useful exception to the habit of technological historians of treating engineers in isolation from other social roles.

4. See A. Michal McMahon, *The Making of a Profession: A Century of Electrical Engineers in America* (New York: IEEE Press, 1984).

5. This point is most clearly made by Monte Calvert, *The Mechanical Engineer in America, 1830–1910; Professional Cultures in Conflict* (Baltimore: Johns Hopkins University Press, 1967). On the history of civil engi-

neering, see Daniel H. Calhoun, *The American Civil Engineer: Origins and Conflict* (Cambridge: Massachusetts Institute of Technology, 1960).

6. For an able discussion of the unfolding meaning of the term *electrician* in the late nineteenth century, see McMahon, pp. 1–29.

7. Alba M. Edwards, *Population: Comparative Occupation Statistics for the United States. 1870–1940. Sixteenth Census of the United States.* (Washington, D.C.: U.S. Government Printing Office, 1943), pp. 146–47.

8. Robert Rosenberg, "Test Men, Experts, Brother Engineers, and Members of the Fraternity: Whence the Early Electrical Work Force?", *Institute of Electrical and Electronics Engineers Transactions on Education* E-17(4) (Nov. 1984): 103–10.

9. C. J. H. Woodbury, "The Barbarians of the Outside World," *Electrical Review*, Apr. 30, 1887, p. 2.

10. *Electrical World*, Feb. 22, 1890, p. 125.

11. Brian Stock, *The Implications of Literacy* (Princeton: Princeton University Press, 1983).

12. *Electrical Review*, Jan. 10, 1891, p. 238.

13. "1882," *Electrician* (London), Jan. 6, 1883, p. 180.

14. "Progress in Wireless Telegraphy," *Electrician* (London), Feb. 9, 1900, p. 552.

15. "Newspaper Science," *Electrician* (London), Nov. 25, 1882, p. 36.

16. Stock, pp. 88–240.

17. Especially see Sylvia Scribner and Michael Cole, *The Psychology of Literacy* (Cambridge: Harvard University Press, 1981), for whom literacy is a "set of organized social practices."

18. *Electrical Review*, Sept. 21, 1889, p. 6.

19. "1885," *Electrician* (London), Jan. 8, 1886, p. 174.

20. "'Electrical' Charlatans," *Electrical Review*, Oct. 22, 1887, p. 2.

21. "A Telegrapher's Troubles," *Electrical Review*, May 23, 1885, p. 7.

22. "Views, News and Interviews," *Electrical Review*, June 6, 1891, p. 195.

23. *Lightning* (London), Oct. 5, 1893, p. 214.

24. "It Would Not Be Blown Out," *Electrical World*, July 11, 1885, p. 18.

25. "Novel Uses of Incandescent Lamps—'A Flash of Darkness'," *Electrical Review*, Oct. 10, 1891, p. 94.

26. *Electrical Review*, Oct. 3, 1891, p. 77.

27. "What Greases the Durn Thing?", *Telegraphic Journal and Electrical Review*, Mar. 13, 1886, p. 6.

28. "A Granger's Experience with the Telephone," *Electrical World*, Oct. 11, 1884, p. 137.

29. *Electrical Review*, June 8, 1889, p. 6.

30. "Railway Travelling Fifty Years Ago," *Answers* (London), July 19, 1890, p. 122.

31. *Western Electrician* (Chicago), June 10, 1899, p. 330.

32. "Explanations That Don't Explain," *Science Siftings* (London), Feb. 2, 1895, p. 213.

33. *Electrical Review,* June 16, 1888, p. 7.

34. "Curiosities of the Wire," *Chambers's Journal* (London), Sept. 2, 1876, p. 566.

35. "Her First Telegram," *Electrical Review,* Mar. 23, 1889, p. 7.

36. *Western Electrician* (Chicago), July 29, 1899, p. 67.

37. "Women as Telegraph Operators," *Electrical World,* June 26, 1886, p. 296. Quoted from the *New York World.*

38. "The Telegraph Message," *Strand* (London), June 1897, p. 691.

39. "They Love the Work," *Electrical World,* Oct. 11, 1884, p. 139. Quoted from the *Somerville Journal.*

40. *Telegraphic Journal* (London), Sept. 1, 1877, p. 208.

41. "Of Course, It Would," *Pittsburgh Chronicle-Telegraph,* reprinted in *Electrical Review,* Nov. 19, 1892, p. 152.

42. "That Telephone Girl Again," *Electrical Review,* Aug. 10, 1889, p. 5. Quoted from the *Albany Journal.*

43. "Women as Telegraph Operators."

44. "That Telephone Girl Again."

45. Ibid.

46. "Lady Phone Operators Should Be Given Elocution Lesson," *Electrical World,* Apr. 11, 1885, p. 141.

47. "Romances of the Telegraph," *Western Electrician* (Chicago), Sept. 5, 1891, p. 130.

48. "Miscellaneous," *Western Electrician* (Chicago), Aug. 29, 1896. Quoted from *Electrical Engineer* (London).

49. *Electrical Review,* Sept. 21, 1889, p. 9.

50. "At the Telephone," *Tit-Bits* (London), Oct. 9, 1897, p. 28.

51. "The Mental and Moral Influence of an Engineering Training," *Electrical World,* Aug. 13, 1898, pp. 158–59.

52. "The Electric Trumpet," *Electrical Review,* Aug. 6, 1887, p. 3.

53. "That Phonograph Again," *Electrical Review,* Sept. 22, 1888, p. 3.

54. "Reducing Salary by Telephone," *Electrical World,* July 11, 1885, p. 18.

55. "Some Electrical Sport," *American Electrician,* Aug. 1898, p. 390.

56. James F. Hobart, "Some Electrical Sport—II," *American Electrician,* Oct. 1897, pp. 408–409.

57. "Electrical Sport," *American Electrician,* Jan. 1898, p. 38.

58. See, for example, "Electrical Sport—Fun with the Boss," *American Electrician,* Feb. 1898, p. 88.

59. *Electrical Review,* May 26, 1888, p. 6.

60. "Abuse of the Telephone," ibid., July 4, 1891, p. 252.

61. "A Bewitched Telegraph Wire," *Scientific American,* May 1, 1880, p. 273.

62. C. J. H. Woodbury, "The Savage and the Circuit," *Electrical Review*, May 16, 1891, p. 160.

63. "Science Against Superstition," *Electrical World*, Aug. 2, 1884, p. 33.

64. Thomas Stevens, "Telegraph Operators in Persia," *Electrical Review*, Aug. 18, 1888, p. 6.

65. *Electrical Review*, Apr. 14, 1888, p. 10.

66. Ibid., Oct. 9, 1985, p. 198. Quoted from the *Jacksonville Times-Union*.

67. *Scientific American*, Feb. 14, 1880, p. 97. Reprinted from the *San Diego Union*.

68. "Apache Chiefs: Indians and the Telephone," *Electrical World*, Aug. 9, 1884, p. 47.

69. *Electrical Review*, Nov. 12, 1892, p. 142.

70. *Lightning* (London), Oct. 20, 1892, p. 247. Quoted from *The Author*.

71. Ibid.

72. "'Electrical' Charlatans," *Electrical Review*, Oct. 22, 1887, p. 2. Reprinted from *The Telegraphist*.

73. "Telegraphic," *Electrical Review*, Oct. 9, 1886, p. 5.

74. *Western Electrician* (Chicago), June 20, 1896, p. 306.

75. *Popular Science News* (Boston), Aug. 1900, p. 144.

76. *Electrical Review*, Oct. 9, 1886, p. 4.

77. *Electrical Review*, Aug. 11, 1895, p. 10.

78. "The Man in the Street on Science," *Electrician* (London), Aug. 20, 1897, p. 546.

79. "Popular Science," *Electrician* (London), May 29, 1891, p. 102.

80. James Payn, "Our Notebook," *Illustrated London News*, July 10, 1897, p. 36.

81. Henry Floy, "The Advisability of Becoming an Electrical Engineer," *Electrical World*, Apr. 14, 1894, pp. 493–94.

82. "Practical Pointers," *Electrical Review*, Nov. 24, 1888, p. 6.

83. *Electrical Review*, Oct. 8, 1887, p. 4. Hibbard had a distinguished telephone career. He became the first general superintendent of AT&T and designed the Blue Bell that became its emblem.

84. *Lightning* (London), June 10, 1897, p. 478.

85. Ibid., Mar. 25, 1897, p. 239.

86. F.L.P., "Marconi in New York," *Western Electrician* (Chicago), Sept. 30, 1899, p. 191.

87. "The Evolution of the Rail Bond," *Electrical World*, Feb. 19, 1898, p. 248.

88. *American Electrician*, Oct. 1898, p. 465.

89. "Interesting Lecture by W. J. Hammer," *Electrical Review*, Mar. 19, 1887, p. 5.

90. *Electrical Review*, July 15, 1896, p. 25.

91. John Kendrick Banks, "The Imp of the Telephone," *Harper's Round Table,* Nov. 19, 1895, p. 67.

92. Henry C. Lahee, "The Minister's Ride," *Comfort* (Augusta, Maine), Feb. 1896, p. 5.

93. Kathryn R. Harris, "Dr. Bogg's Experiment," *Comfort* (Augusta, Maine), Apr. 1900, pp. 4–5.

94. "Electric Fruit," *Electrical Review,* Nov. 3, 1888, p. 1.

95. *Electrical Review,* Oct. 8, 1887, p. 6.

96. Ibid., Feb. 4, 1888, p. 6.

97. *Western Electrician* (Chicago), Aug. 16, 1890, p. 87. Quoted from the *San Francisco Chronicle.*

98. John Coleman Adams, "Democracy and the Mother Tongue," *Cosmopolitan,* Feb. 1893, p. 461.

99. *Electrical World,* Jan. 14, 1885, p. 61.

100. "New Electrical Dictionary," *Western Electrician* (Chicago), May 28, 1892, p. 323.

101. *Electrical Review,* Oct. 8, 1887, p. 7.

102. E. Hospitalier, "Nomenclature, Symbols and Notation," *Electrical World,* Aug. 26, 1893, p. 144.

103. "Electrical Notation, Symbols and Abbreviations," *Electrical World,* Sept. 23, 1893, p. 225.

102. "Which Shall It Be?", *Electrical Review,* Aug. 17, 1889, p. 20.

105. "Electrical Nomenclature and Notation," *Electrical World,* June 27, 1885, p. 252.

106. George W. Mansfield, "Shall It Be 'Motoneer?'," *Electrical World,* Nov. 15, 1884, p. 198.

107. Thomas D. Lockwood, "A Mutineer Against Motoneer," *Electrical World,* Nov. 21, 1884, p. 211.

108. "Ampere," "It Must Not Be Motoneer," *Electrical World,* Nov. 29, 1884, p. 233.

109. "The Extension of Sense," *Scientific American,* Dec. 16, 1876, p. 384.

110. "The Convention of the National Electric Light Association at New York, May 5–7, 1896," *Western Electrician* (Chicago), May 16, 1896, p. 238.

111. *Electrical Review,* Aug. 20, 1887, p. 6.

112. William S. Hine, "Electrical Nomenclature," reprinted ibid., Apr. 30, 1887, p. 2.

113. William Crookes, "Some Possibilities of Electricity," *Fortnightly Review,* Feb. 1892, p. 179.

114. "The Convention of the National Electric Light Association," *Western Electrician* (Chicago), May 16, 1896, p. 238.

115. Julian A. Moses, "Electricity and Conjuring," *Electrical Review,* Nov. 19, 1892, p. 152.

116. *Electrical Review,* Mar. 19, 1887, p. 5.

117. *Electrical World*, Aug. 4, 1894, p. 97.

118. "Hertzian Waves at the Royal Institution," *Electrician* (London), June 8, 1894, p. 156.

119. Nevil Monroe Hopkins, "Electrical Apparatus for Amusement—III. The Spiritualistic Cash Box," *American Electrician*, Mar. 1898, p. 117; Woodbury, "The Savage and the Circuit," p. 160.

120. *Electrical Review*, Feb. 27, 1892, p. 1.

121. "A Train Dispatcher's Story," ibid., Nov. 6, 1886, pp. 6–7.

122. "Röntgen-Ray Ghosts in Paris," *Popular Science News* (Boston), May 1897, p. 110.

123. "The Immediate Future of the Long-Distance Telephone," *Current Literature*, May 1911, p. 504.

124. Ibid.

2 Community and Class Order

1. See "Telephones and Average Daily Conversations (Bell and Independent Companies): 1876 to 1900," Series R 1–12, in U.S. Bureau of Census, *Historical Statistics of the United States*, Part 2 (Washington, D.C., 1975), pp. 783–84. See also "The Delegates to the Telephone Convention," *Electrical Review*, Sept. 21, 1889, p. 8. See also Claude Fischer, "Technology's Retreat: The Decline of Rural Telephony, 1920–1940" (scheduled for publication in *Social Science History*, Summer 1987).

2. "The Future of the Telephone," *Scientific American*, Jan. 10, 1880, p. 16.

3. "Epoch-Making Inventions of America," *Electrical Review*, Apr. 18, 1891, p. 109.

4. "Latest Wonders of the Phonograph," *Invention* (London), Apr. 30, 1898, p. 278.

5. "London's Visitors a Day," *Answers* (London), Nov. 10, 1894, p. 445.

6. Michael Schudson, *Discovering the News* (New York: Basic Books, 1978), pp. 59–60.

7. "Telephone Cranks," *Western Electrician* (Chicago), July 17, 1897, p. 37.

8. "He Wasn't John L. Sullivan," *Electrical Review*, July 27, 1889, p. 8. Quoted from the *Evening Sun*.

9. *Harper's*, Sept. 1893, p. 726.

10. "The Dead Alive in the Biograph," *Popular Science News* (Boston), Oct. 1901, p. 233.

11. "The Telephone Unmasked," *Telegraphic Journal* (London), Dec. 1, 1877, p. 292. Quoted from the *New York Times*.

12. *American Electrician*, Jan. 1897, p. 15.

13. "How Telephone Lines Are Worked in Scandal Cases," *Electrical Review*, Apr. 13, 1899, p. 6, and "Views, News and Interviews," *Electrical Review*, Oct. 17, 1891, p. 107.

14. Quoted in "Telephonic Conversation as Evidence," *Electrical Review*, Dec. 4, 1886, p. 9.

15. "Secrecy of the Telephone," *Electrical Review*, Feb. 9, 1889, p. 6. See also "Secrecy of Telegraphic Dispatches," *Western Electrician* (Chicago), Mar. 28, 1896, p. 151.

16. "Kitchen Chats," *Comfort* (Augusta, Maine), Aug. 1896, p. 7. See also "Looking Forward," *Siftings* (London), May 6, 1893, p. 41, and the Marquis of Salisbury, speech to the first annual dinner of the Institute of Electrical Engineers, London, *Electrician*, Nov. 8, 1889, p. 13, for similar opinions.

17. "Trespass by Telephone," *Telephony*, (Chicago), Sept. 1905, p. 221.

18. "Elocution over the Telephone," *Electrical Review*, Nov. 27, 1886, p. 6.

19. "Where Telegrams Are Rare," *Electrical World*, Jan. 30, 1886, p. 48.

20. *Western Electrician* (Chicago), July 29, 1899, p. 67.

21. "The Telephone Girl Again," *Electrical Review*, Aug. 10, 1889, p. 6.

22. "The Matinee Girl," *Munsey's*, Oct. 1897, p. 34.

23. Quoted in "The Ticker Wasn't Tickled," *Electrical Review*, Nov. 26, 1892, p. 167.

24. "Enterprising Telephone Suitor," *Electrical Review*, Dec. 14, 1889, p. 4.

25. *Electrical Review*, Mar. 3, 1888, p. 11. (Quoted from the *New York Sun*.)

26. "A Bad Phonograph Under the Sofa," ibid., Sept. 29, 1888, p. 7.

27. "The Dangers of Wired Love," *Electrical World*, Feb. 13, 1886, pp. 68–69.

28. "So Far and Yet So Near," *Electrical World*, Mar. 1, 1884, p. 68.

29. Ibid.

30. *Electrical Review*, June 15, 1889, p. 6.

31. "Healthiness of the Electric Light," ibid., Jan. 19, 1889, p. 6.

32. *Lightning* (London), April 21, 1892, p. 591.

33. "Electricity in the Household," *Invention* (London), Dec. 17, 1898, p. 811.

34. "Burglars and Electric Alarums," *Electrician* (London), Sept. 29, 1882, p. 469.

35. "The St. Louis Anticipation," *Electrical Review*, Jan. 21, 1888, p. 6.

36. "Electrical Installation at Greenwich, Ct.," *Electrical Review*, Oct. 22, 1887, p. 1.

37. "Which One of These Suit You? Ideal Houses with More Than the Latest Improvements," *Answers* (London), Jan. 6, 1893, p. 118.

38. "Queries and Answers," *Science Siftings* (London), Mar. 24, 1900, p. 326.

39. "The Improved Gramophone," address by Emile Berliner to the Institute of Electrical Engineers, Dec. 16, 1890. Reprinted in *Western Electrician* (Chicago), Jan. 3, 1891, p. 5.

40. "Extension and Improvement of the Telephone Service," *Electrical World,* Sept. 20, 1890, p. 197.

41. "Music on Tap," *Electrical World,* Sept. 20, 1890, p. 195.

42. *Electrical Review,* Mar. 12, 1887, p. 5.

43. "Diseased Germs Transmitted Through Telephone Circuits," *Electrical World,* June 23, 1894, p. 833.

44. "Queries and Answers," *Science Siftings* (London), June 11, 1898.

45. "A Grand Invention," *Electrical Review,* Aug. 21, 1886, p. 5. Quoted from *Court Journal.*

46. *The Million* (London), Sept. 10, 1892, p. 255.

47. See also "This Clock Shouts at You," *Answers* (London), Feb. 25, 1898, p. 278, and "New Electric Alarm Clock," *Popular Science News* (Boston), Jan. 1901, p. 13.

48. *The Million* (London), Dec. 31, 1892, p. 263.

49. "Useful in the Household," *Answers* (London), Apr. 30, 1892, p. 399.

50. *Science Siftings* (London), Sept. 2, 1899, p. 287. For a description of the mechanism of the talking clock, see "Clocks Which Speak the Time," *Answers* (London), Feb. 11, 1893, p. 205.

51. "Need 'Hello Girls' as Alarm Clocks," *Telephony* (Chicago), Dec. 1905, p. 415.

52. *Electrical Review,* Mar. 10, 1888, p. 9.

53. "Need 'Hello Girls' as Alarm Clocks."

54. "Said She Liked It," *Electrical Review,* Nov. 14, 1891, p. 168.

55. "Need 'Hello Girls' as Alarm Clocks."

56. "Said She Liked It."

57. "Danger in Telephoning Society Leaders," *Western Electrician* (Chicago), June 10, 1899, p. 330.

58. "A Queen Who Believes in Progress," *Electrical Review* (London), July 8, 1896, p. 15.

59. "Yankees Telephone King Edward," *Telephony* (Chicago), Aug. 1905, p. 112.

60. "Telephone Cranks," p. 37.

61. "Telephoning is the National Craze," *Telephony* (Chicago), Dec. 1905, p. 412.

62. Remarks by W. H. Eustis at the National Telephone Exchange Association banquet, Sept. 12, 1889, reported in "The Delegates to the Tele-

phone Convention Entertained by the Erie Officials—The Toasts," *Electrical Review*, Sept. 21, 1889, p. 9.

63. *Electrical Review*, Oct. 20, 1888, p. 6.

64. "Telephone Cranks."

65. *Electrician* (London), Mar. 20, 1886, p. 6. Quoted from the *Leavenworth Times*.

66. "To What Base Uses," *Electrical World*, Nov. 29, 1884, p. 222.

67. "The Telephone in Conversation," *Electrical Review*, Feb. 23, 1889, p. 4.

68. "Swearing by Telephone," *Scientific American*, Apr. 7, 1883, p. 209.

69. Remarks by W. H. Eustis, National Telephone Exchange Association.

70. *Electrical World*, Mar. 1, 1884, p. 68.

71. Quoted in "Use of the Telephone," *Western Electrician* (Chicago), Oct. 24, 1891, p. 247.

72. "Stories of the Telephone," *Western Electrician* (Chicago), June 10, 1899, p. 330. Quoted from the *New York Tribune*.

73. Ibid.

74. Ibid.

75. "After-Dinner Speeches by Telephone," *Telephony* (Chicago), Aug. 17, 1912, p. 46.

76. See Jeffrey Kieve, *The Electric Telegraph: A Social and Economic History* (London: David & Charles, 1973), p. 39.

77. Harold Scott, ed., "Crippen, Hawley Harvey," *The Concise Encyclopedia of Crime and Criminals* (New York: Hawthorn Books, 1961), p. 109.

78. "Heavy Swindling by Telephone," *Electrical Review*, Aug. 11, 1888, p. 8. Quoted from the *Chicago Herald*.

79. *Electrical World*, June 19, 1886, p. 28. See also "Mr. W. L. Carpenter on Daily Practical Applications of Electricity in America," *Telegraphic Journal* (London), Dec. 1, 1879, pp. 385–86.

80. *Electrical World*, June 19, 1886, p. 28.

81. "A Marriage by Telegraph Annulled," ibid., Nov. 21, 1884, p. 211.

82. *Electrical World*, Feb. 14, 1885, p. 61.

83. "Romances of the Telegraph," *Western Electrician* (Chicago), Sept. 5, 1891, pp. 130–31. Quoted from the *San Diego Union*.

84. "Swindling by Telephone," *Electrical Review*, Oct. 15, 1887, p. 6.

85. "The Telephone and the Man With a Thirst," *Electrical Review*, Sept. 12, 1891, p. 30.

86. "Views, News and Interviews," *Electrical Review*, May 2, 1891, p. 133.

87. "To What Base Uses."

88. Allen Grant, "An Electric 'Scoop,'" *Comfort* (Augusta, Maine), May 1898, pp. 4–5.

89. "Burglary on Party Line," *Telephony* (Chicago), Jan. 1906, pp. 54–55.

90. *Electrical Review*, Apr. 2, 1892, p. 77.

91. Ibid., Feb. 27, 1892, p. 1.

92. Ibid., Apr. 2, 1892, p. 81.

93. "Secret Photography to Crack Crimes," ibid., Sept. 22, 1888, p. 3.

94. "The New Phonograph," *Scientific American,* Dec. 31, 1887, p. 422.

95. "Telegraphing from the Grave. And Some Other Startling Possibilities," *Answers* (London), May 27, 1899, p. 48.

96. *Telegraphic Journal* (London), May 15, 1878, p. 208.

97. *Electrical Review*, Aug. 25, 1888, p. 7.

98. Ibid., Feb. 27, 1892, p. 5.

99. Speech by C. J. H. Woodbury, "The Barbarians of the Outside World," New York Electric Club, Apr. 21, 1887, reprinted in "The Electric Club's Annual," *Electrical Review*, Apr. 30, 1887, p. 2.

100. "Experimental Messages Which Flashed Through Albany," *Electrical Review*, Feb. 11, 1888, p. 1.

101. *Electrical Review* (London), June 15, 1889, p. 6.

102. In *Democracy and Liberty,* quoted in *Electrician* (London), June 5, 1896, p. 166.

103. "Prince Bismarck's Lament," *Electrical World*, Jan. 3, 1885, p. 1.

104. Ibid.

105. "The New Orleans Exposition," *Electrical World*, Feb. 20, 1884, p. 257.

106. "Views, News and Interviews," *Electrical Review*, Aug. 27, 1891, p. 1.

107. "Telephones for the London Police," *Electrical Review*, July 7, 1888, p. 4. Quoted from the *Electrician*.

108. *Telegraphic Journal* (London), Dec. 1, 1879, p. 386.

109. "Electricity as Executioner," *Telegraphic Journal* (London), Jan. 15, 1876, p. 32. Quoted from *Scientific American*.

110. "1993. Glimpses of 100 Years Ahead," *Answers* (London), Feb. 4, 1893, p. 190.

111. *Lightning* (London), Jan. 5, 1893, p. 1.

112. *Electrical World*, July 19, 1884, p. 19.

113. *Lightning* (London), June 27, 1895, p. 447. Quoted from the *Daily Chronicle*, Mar. 2, 1895.

114. "The Telephone," *Electrical Review*, Sept. 15, 1888, p. 15.

115. *Electrical World*, Sept. 26, 1885, p. 127. Quoted from the *New York World*.

116. *Electrician* (London), Dec. 27, 1889, p. 185.

117. *Proceedings of the Third Meeting of the National Telephone Exchange Association* (New York: American Bell Telephone Company, 1881), p. 82. I am indebted to Milton Mueller for this example.

118. Ibid.

119. "'Not Quite So Free, Please,'" *Electrical World*, Jan. 3, 1885, p. 8. Quoted from the *Scotsman*, Dec. 9, 1884.

120. "Telephone Cranks."

121. *Electrical Review*, June 16, 1888, p. 7.

122. *Electrical World*, Apr. 11, 1885, p. 141.

123. *Electrical Review*, Nov. 2, 1889, p. 1.

124. *Electrical World*, July 4, 1885, p. 1.

125. "Telephone War in Washington," *Western Electrician* (Chicago), Feb. 19, 1898, p. 109.

126. See also "Removal of Police Telephones in Chicago," *Western Electrician* (Chicago), Feb. 1, 1890, p. 53, which reports that the Chicago Telephone Company removed fifteen telephones from various city departments a year after installing them because there was no appropriation to pay the bills that had come due. Further: "It appears that these instruments . . . were not used for official business exclusively, in fact, this formed a very small portion of the service. . . . The instruments that were removed were used chiefly by the newspaper reporters and persons in the vicinity of the police stations."

127. *Electrical Review*, Mar. 7, 1891, p. 43.

128. "The Drug Store System," ibid., Oct. 20, 1888, p. 7. Quoted from the *Evening Journal*.

129. "A New Telephone Idea," *Popular Science Monthly*, May 1897, p. 156. This article reports that in 1897 "Washington druggists" paid $100 to $125 for the telephone. Those who paid a special rate of $140 were permitted to offer free use of the telephone to their patrons.

130. *Western Electrician* (Chicago), July 29, 1899, p. 76.

131. *Electrical Review*, July 20, 1889, p. 4.

132. *Western Electrician* (Chicago), Jan. 20, 1900, p. 42.

133. *Electrical World*, Jan. 24, 1885, p. 38.

3 Locating the Body in Electrical Space and Time

1. "Popular Science," *Electrician* (London), May 29, 1891, p. 102.

2. "Telegraphy Puzzled Him," *Electrical Review*, Mar. 23, 1889, p. 7.

3. "What is Electricity?", *Telephony* (Chicago), Sept. 1905, p. 221. Quoted from *Popular Science Monthly*.

4. Classic statements of this presumed division are found in, among others, Jack Goody, *The Domestication of the Savage Mind* (Cambridge: Cambridge University Press, 1977). For a critique of this dichotomy, see Carolyn Marvin, "Constructed and Reconstructed Discourse: Inscription and Talk in the History of Literacy," *Communications Research* 11 (Oct. 1984): 563–94, and Harvey Graff, "Reflections on the History of Literacy," *Humanities in Society*, Fall 1981, pp. 391–92.

5. Quoted in Park Benjamin, "The Possibilities of Electricity," *Forum*, Dec. 1889, pp. 391–92.

6. William Crookes, "Some Possibilities of Electricity," *Fortnightly Review* (London), Feb. 1892, p. 173.

7. Ibid., p. 176.

8. Ibid., pp. 180–81.

9. Henry Morton, "The Dangers of Electricity," speech to the twelfth convention of the Electric Light Association, held Aug. 19–21, 1890, reported in *Western Electrician* (Chicago), Aug. 30, 1890, p. 105.

10. "Stories of the Telephone," *Western Electrician* (Chicago), June 10, 1899, p. 330. Quoted from the *New York Tribune*.

11. "Lion Taming by Electricity," *Electrical World*, Sept. 27, 1884, p. 108. Quoted from the *Pall Mall Gazette*.

12. "Electricity for a Balky Horse," *Popular Science News* (Boston), Sept. 1897, p. 207.

13. "Electricity in the Jungle," *Western Electrician* (Chicago), July 9, 1892, p. 18. Quoted from the *North American Review*.

14. "The Spider and Its Telephone Wires," *Electrical World*, June 26, 1886, p. 296.

15. "The Electric Light," *Electrical World*, Oct. 18, 1884, p. 151.

16. "The Electric Light as an Insect Destroyer," *Electrical World*, June 6, 1885, p. 229.

17. These are found in James F. Hobart, "Some Electrical Sport—III," *American Electrician*, Nov. 1897, p. 452; "Some Electrical Sport—IV," Dec. 1897, p. 489; and Andrew J. Rodgers, Letters to the Editor, "More Electrical Sport," Mar. 1898, p. 127.

18. "Electric Lights for a Bull Fight," *Electrical Review*, May 7, 1887, p. 8.

19. *Electrical Review*, July 10, 1886, p. 9.

20. Ibid., Apr. 23, 1887, p. 9.

21. Reported in "Miscellaneous," *Western Electrician* (Chicago), Aug. 29, 1896, n.p.

22. *Telegraphic Journal* (London), Sept. 1, 1877, p. 208.

23. *Popular Science News* (Boston), Sept. 1897, p. 207.

24. *Electrical Review*, Nov. 12, 1892, p. 142.

25. *Western Electrician* (Chicago), July 19, 1890, p. 39.

26. *Electrical Review*, Aug. 18, 1888, p. 4. Quoted from the *Milwaukee Sentinel*.

27. Ibid., Oct. 9, 1886, p. 4.

28. Ibid., Mar. 24, 1888, p. 2. Quoted scornfully from the *New York World*.

29. "A Western Scientist," ibid., Feb. 2, 1889, p. 3.

30. *Electrical World*, June 5, 1886, p. 257.

31. *Electrical Review*, Oct. 9, 1886, p. 4.

32. Ibid.

33. *Lightning* (London), Dec. 22, 1892, p. 39.

34. See "Recovery from Alleged Severe Electric Shock," *Electrical World,* June 27, 1885, p. 252, quoting from an article by Dr. A. L. Hummel in *Medical Bulletin.*

35. *Electricity* (London), Dec. 27, 1899, p. 386.

36. *Popular Science News* (Boston), Apr. 1897, p. 87.

37. "Views, News, and Interviews," *Electrical Review,* Oct. 10, 1891, p. 1.

38. See, for example, "Electrical Initiations," *Popular Science News* (Boston), Jan. 1899, p. 14; *Western Electrician* (Chicago), Nov. 26, 1898, p. 300; ibid., Dec. 9, 1899, p. 10; same story in *Electricity* (London), Dec. 27, 1899, p. 386; *Telegraphic Journal* (London), Sept. 1, 1877, p. 209; "Survived an Electric Shock," *Popular Science News* (Boston), June 1897, p. 135; *Electrical Review,* Jan. 26, 1894, p. 4.

39. "An Electrified Bridge," *Popular Science News* (Boston), Sept. 1897, p. 207.

40. *Telegraphic Journal* (London), Sept. 1, 1877, p. 208.

41. *Electrical Review,* Aug. 8, 1885, p. 6.

42. Ibid., Sept. 24, 1887, p. 4.

43. "Exhibits of Electrical Interest at the Mechanic Fair, Boston," *Electrical World,* Oct. 29, 1898, p. 451.

44. "Electric Jewelry," *Electrical World,* Aug. 9, 1884, p. 43.

45. John Kendrick Bangs, "The Imp of the Telephone," *Harper's Round Table,* Nov. 19, 1895, p. 67.

46. "New Dining Room Push Button," *Western Electrician* (Chicago), Oct. 11, 1890, p. 195.

47. "The Buttoned Kid. A Tandem Escapade," *Yellow Kid,* May 22, 1897, p. 3.

48. "Looking Backward," *Electrician* (London), Feb. 11, 1892, p. 366.

49. "Objection to the Telephone as a Missionary," *Electrical Review,* 14(21):6, quoted from *Christian at Work* (London); also, *Electrical Review,* June 15, 1889, p. 6, quoted from *The Telephone.*

50. "Faraday's Religious Life," *Western Electrician* (Chicago), July 18, 1891, p. 36.

51. "An Electrical Dedication," *Electrical Review* (London), June 24, 1896, p. 138.

52. Quoted from *Messiah's Herald* in D. T. Taylor, *A Chariot of Fire. The Cars in Prophecy and History with the Wonders of Rapid Traveling and Significance of the Modern Railway System. A Token of the Nearing End of the Age* (Yarmouth, Maine: Scriptural Publication Society, 1888), p. 160.

53. Edward C. Towne, "Magnetism and Electricity," *Science Siftings* (London), May 13, 1893, p. 25.

54. Erastus Wiman, "Electrical Possibilities as Viewed by a Business Man," *Electrical Review,* June 15, 1889, p. 3.

55. "An Electrical Dedication."

56. Heinrich Hertz, "Identity of Light and Electricity," Sept. 1889, reported in *Electrical Review*, Dec. 14, 1889, p. 7.

57. Wiman, "Electrical Possibilities."

58. "Future of Electrical Development," interview with Amos Dolbear, *Western Electrician* (Chicago), Jan. 9, 1897, p. 24.

59. *Western Electrician* (Chicago), May 8, 1886, p. 7; "An Electrical Grave-Annunciator," ibid., July 18, 1891, p. 35. See also "To Prevent Untimely Burial," *Electrical Review*, Nov. 6, 1886, p. 1.

60. "Telegraphing from the Grave. And Some Other Startling Possibilities," *Answers* (London), May 27, 1899, p. 48.

61. The translated work was "A Ternary of Paradoxes of the Magnetic Cure of Wounds, etc.," Walter Charleton, trans., 2nd ed. (London, 1650). Quoted in Fred de Land, "Notes on the Evolution of Electrical Transmission of Speech," *Telephony* (Chicago), Dec. 1901, p. 212.

62. *Electrical Review*, Apr. 19, 1899, p. 227. Quoted from the *Electrical Review* of London.

63. *Electrical Review*, Oct. 6, 1888, p. 9.

64. "Electricity in Your Mouth," ibid., June 1, 1889, p. 10.

65. "An Aid to the Surgeon," *Electrical Review*, Aug. 10, 1889, p. 1.

66. "Therapeutic Effects of High Frequency Currents," *Western Electrician* (Chicago), Sept. 12, 1896, p. 123.

67. *Western Electrician* (Chicago), Dec. 16, 1899, p. 354.

68. "Future of Electrical Development."

69. Crookes, "Some Possibilities of Electricity," pp. 179–80.

70. "Electricity Finds a Place in Fiction," *Electrical Review*, May 5, 1888, p. 7.

71. "Miscellaneous," *Western Electrician* (Chicago), Aug. 29, 1897, n.p.

72. "An Electrical Hat," *Electrical Review*, Oct. 23, 1886, p. 3.

73. *Electricity* (London), June 26, 1896, p. 307.

74. "More Electrical Quackery," *Science Siftings* (London), Sept. 26, 1896, p. 365.

75. "An Electrical Treat," *Electrical Review*, Feb. 26, 1887, p. 13.

76. Louis E. Asher and Edith Neal, *Send No Money* (Chicago: Argus Books, 1942), pp. 62–63.

77. See "Electric Belts in England," *Electrical World*, Sept. 1, 1894, p. 207.

78. *Answers* (London), June 1, 1895, p. 201.

79. "Views, News, and Interviews," *Electrical Review*, June 6, 1891, p. 195.

80. "The Wizard of the West," advertisement in *Comfort* (Augusta, Maine), Nov. 1899, p. 24.

81. "The Telephone as a Cause of Ear Troubles," *Electrical Review*, Oct. 26, 1889, p. 6. Quoted from the *British Medical Journal*.

82. *Western Electrician* (Chicago), Sept. 13, 1890, p. 147. Another "man who imagined himself a telephone, and who has been trying for a year to

shout 'hello!' in his own ear, has been sent to an asylum at Flatbush, Long Island," reported the *Electrical Review*, Feb. 5, 1887, p. 8.

83. "Troubled by Telephone Bells," *Electrical Review*, Mar. 14, 1891, p. 54.

84. "An Electric Crank," *Electrical Review*, Oct. 24, 1891, p. 126.

85. "Prince Bismarck's Lament," *Electrical World*, Apr. 12, 1884, p. 120.

86. *Electrical Review*, Oct. 30, 1886, p. 5.

87. "A Very Good Story of a Living Magnet," ibid., June 9, 1888, p. 11.

88. "Heart Inverted by Lightning," *Popular Science News* (Boston), Nov. 19, 1901, p. 253. Quoted from *Electrical Review*.

89. "An Invention Worthy of Munchausen," *Electrical Review*, Aug. 29, 1891, p. 9.

90. *Electrical Review*, July 6, 1889, p. 15.

91. Fernand Papillon, "Electricity and Life," *Popular Science Monthly*, Mar. 1873, p. 529. Translated by A. R. Macdonough from the *Revue des Deux Mondes*.

92. "The Freaks of Lightning," *Popular Science News* (Boston), July 1901, p. 156.

93. "The Electric Boy," *Electrical Review*, Oct. 8, 1887, p. 7.

94. Quoted by *Lightning* (London), July 6, 1893, pp. 1–2. See also "A New Electric Light," *Popular Science Monthly*, May 1892, p. 137.

95. "Hertzian Waves at the Royal Institution," *Electrician* (London), June 8, 1894, p. 156.

96. "The Use of Illuminated Girls," *Electrical World*, May 10, 1884, p. 151.

97. *Electrical Review*, Feb. 7, 1885, p. 4.

98. "An Electric Supper," *Electrical World*, Feb. 6, 1886, p. 56.

99. "Electricity in Iowa," *Western Electrician* (Chicago), June 27, 1891, p. 367. Quoted from the *Sheldon* (Iowa) *Eagle*.

100. "The Representative of the Electric Light," *Western Electrician* (Chicago), Apr. 12, 1890, p. 210.

101. "Trouvé's Jewelry," *Electrical Review*, June 27, 1885, p. 2. The earliest description of this jewelry I have found is "Electric Jewelry," *Scientific American*, Oct. 25, 1879, p. 263, quoted from *La Nature*.

102. "A Paris Electric Bell(e)," *Electrical World*, Nov. 29, 1884, p. 223.

103. Thomas Stevens, "Telegraph Operators in Persia," *Electrical Review*, Aug. 18, 1888, p. 7.

104. Printed in the *Electrician* (London), Sept. 20, 1895, p. 689. See also "Wireless Telegraphy," *Cram's Magazine* (Chicago), Jan. 1900, p. 229, for another example of the use of organic-electric metaphor.

105. Forée Bain, "Looking Forward," *Western Electrician* (Chicago), Jan. 5, 1901, p. 11.

106. "The Doll with a Memory," *Literary Digest,* Jan. 2, 1897, pp. 271–72. Quoted from a paper by Reginald A. Fessenden in the *Journal of the Franklin Institute.*

107. "An Electric Supper."

108. *Electrical Review,* May 2, 1891, p. 138. Quoted from the *Chicago News.*

109. "A Bridgeport Inventor Builds a Boy Who Walks and Smiles," *Electrical Review.* The Xerox reproduction of this source in my possession unfortunately does not include the date.

110. Ibid.

111. E. I. La Baueme, "Visions of 1950," *Technical World* (Chicago), Dec. 1911, p. 439.

112. *Western Electrician* (Chicago), Dec. 6, 1890, p. 308. Quoted from the *Electrical Review* (London).

113. Remarks by J. J. Davenport, the mayor of Kansas City, at the eleventh convention of the National Electric Light Conference, Kansas City, Feb. 11, 1890, quoted in *Electrical World,* Feb. 22, 1890, p. 125. See also "Electricity in the Navy," *Electrical Review,* June 27, 1885, p. 10.

114. "Electrical Wiring in Harbors," *Western Electrician* (Chicago), Apr. 16, 1898, p. 227.

115. *Telephony* (Chicago), Sept. 1905, p. 221.

116. "Electric Shells in Warfare!", *Electrical Review,* Feb. 16, 1889, p. 4.

117. Quoted ibid.

118. "Edison Out-Edisoned," *Lightning* (London), Feb. 6, 1896, p. 111.

119. "Oh Dear! Another War Terror," *Answers* (London), July 30, 1897, p. 182. Marconi specifically denied that he had made such predictions ("Wireless Telegraphy. Signor Marconi Speaks of Its Value," *Daily Chronicle,* Sept. 20, 1897).

120. *Electrical Review* (London), Mar. 29, 1899, p. 179; "Electricity's Mystery," *Western Electrician* (Chicago), Dec. 10, 1898, p. 332; "A Terrible War Prospect," *Science Siftings* (London), June 25, 1898, p. 142; *Electrician* (London), May 26, 1899, p. 144.

121. "In 1992. Five Minutes with the Future," *Answers* (London), Sept. 3, 1892, p. 258.

122. "Electricity in Warfare," *Western Electrician* (Chicago), Apr. 18, 1891, p. 221.

123. J. J. Woods, "Strange Adventures in the Year 2890; or, A Thousand Years into Futurity," *Answers* (London), Jan. 4, 1890, p. 96.

124. Edward Bulwer-Lytton, *The Coming Race* (London: George Routledge and Sons, 1871), p. 64.

125. Ibid., p. 56.

126. Nikola Tesla, "The Problems of Increasing Human Energy," *Century,* June 1900, p. 183.

127. Ibid.

128. Hugh Tuite, "For Valour. A War Story of the Twentieth Century," *Daily Mail* (London), Oct. 24, 1898, "Daily Magazine" column.

129. See, for example, Charles R. Huntley, "The Execution as Seen by an Electrician," *Electrical World*, Aug. 16, 1890, p. 100.

130. Quoted in "Kemmler's Terrible Death," *Western Electrician* (Chicago), Aug. 16, 1890, p. 84.

131. Quoted ibid.

132. This was the position taken by George Washington, editor of the *Electrical Review*, and reflected in many of its articles. By far the most common argument in the debate concerned whether electrocution could be administered without unusual cruelty and suffering to those marked for punishment. Some argued for electrocution on the grounds that it was more humane than the rope; others countered that if it were indeed more humane, the deterrent value of capital punishment would be diminished.

133. Quoted in "Kemmler's Terrible Death."

134. "Prof. Brackett Opposed to Executions by Electricity," *Electrical Review*, Dec. 28, 1888, p. 4.

135. "The Electrical Execution," *Electrical Review*, Nov. 24, 1888, p. 6. Quoted from "The Electrical Execution," *New York World*, Nov. 16, 1888.

136. Matthew Josephson, *Edison* (New York: McGraw-Hill, 1959), pp. 347–49. See "Postponement of Kemmler's Execution," *Western Electrician* (Chicago), May 10, 1890, for the text of a public statement by George Westinghouse.

137. "Electricity as an Executioner," *Telegraphic Journal* (London), Jan. 5, 1876, p. 32. Quoted from *Scientific American*.

138. *Electrical World*, Aug. 16, 1890, p. 100.

4 Dazzling the Multitude

1. Erik Barnouw has traced a similar theme concerning the magic spectacle as a predecessor of certain elements of cinema in *The Magician and the Cinema* (New York: Oxford University Press, 1981).

2. *Electrician* (London), Oct. 22, 1897, p. 847. Quoted from the *Dundee Advertiser*.

3. Quoted in Fred de Land, "Notes on the Evolution of Electrical Transmission of Speech," *Telephony* (Chicago), Dec. 1901, p. 215.

4. "Edison's Telepathic Machine," *Science Siftings* (London), Dec. 22, 1894, p. 129. R. Famianus Strada, *Prolusiones Academicas* (Rome, 1617; Oxford, 1662). See *Telegraphic Journal* (London), Nov. 15, 1875, pp. 256–57, for the relevant text in Italian and translation from the Oxford edition.

5. "The Electrical Exhibition at Paris," *Electrician* (London), Dec. 3, 1881, pp. 40–41.

6. *Invention* (London), July 24, 1897, p. 55.

7. Houghton Townley, "The Long, Long Sleep," *Answers* (London), Aug. 12, 1899, pp. 244–46.

8. William Crookes, "Some Possibilities of Electricity," *Fortnightly Review* (London), Feb. 1892, pp. 174–75.

9. Ayrton's lecture, "Sixty Years of Submarine Telegraphy," was delivered Feb. 15, 1897, at the Imperial Institute and printed in the *Electrician* (London), Feb. 19, 1897, p. 548.

10. Remarks by E. J. Hall, Jr., *Electrical Review*, Sept. 21, 1889, p. 9.

11. Marshall McLuhan, *Understanding Media: The Extensions of Man* (New York: McGraw-Hill, 1964), p. 9.

12. "Electrical Illumination at the Peace Jubilee," *Western Electrician* (Chicago), Oct. 29, 1898, p. 39.

13. "Illumination at the Knights Templar Conclave," *Electrical World*, Oct. 22, 1898, pp. 411–12.

14. "Lighting the Statue of Liberty," *Electrical World*, Apr. 26, 1884, p. 136.

15. *Electrical Review*, Oct. 16, 1886, p. 9.

16. "The Electrical Exhibition at Paris," *Electrician* (London), Dec. 3, 1881, p. 40.

17. "The Paris Exposition," *Western Electrician* (Chicago), Jan. 6, 1900, p. 1.

18. "The Electric Light Considered Irreligious," *Electrical World*, Jan. 24, 1885, p. 32, and "Our London Letter," *Electrical Review*, Aug. 27, 1892, p. 7.

19. "Christmas and Electricity," *Electrical World*, Dec. 22, 1894, p. 651.

20. *Electrical World*, Apr. 11, 1885, p. 141; "1884," *Electrician* (London), Jan. 3, 1885, p. 157.

21. Bejohn W. Beane, "The History and Progress of Electric Lighting," *Proceedings of the National Electric Light Association* (Baltimore: Baltimore Publishing Co., 1886), p. 35.

22. Ibid., p. 35.

23. *Electrician* (London), Jan. 5, 1884, p. 183.

24. *Electrical Review*, Nov. 20, 1886, p. 4. Quoted from the *New York Tribune*.

25. "1881," *Electrician* (London), Jan. 7, 1882, p. 120, and Beane, pp. 29–30.

26. "A Town Lighted by Electricity," *Scientific American*, May 1, 1880, p. 275.

27. Ralph Hower, *History of Macy's of New York: 1858–1919* (Cambridge: Harvard University Press, 1943), pp. 166, 167, 283, 450.

28. Jefferson Williamson, *The American Hotel: An Anecdotal History* 'ew York: Alfred A. Knopf, 1930), p. 66.

29. Quoted in "Electrically Lighted Signs," *Electrical Review,* May 8, 1895, p. 249.

30. *Telegraphic Journal* (London), Jan. 1, 1881, p. 17.

31. "Mr. Edison as an Electrical Litterateur," *Electrical World,* Jan. 24, 1885, p. 38. Edison quote from the *Boston Herald,* Jan. 18, 1885.

32. Beane *Proceedings,* pp. 38–39.

33. *Proceedings of the Ninth National Electric Light Association Conference,* Feb. 1889; address by President M. J. Francisco at the eighteenth National Electric Light Association Conference, Feb. 1895.

34. *Electrician* (London), Jan. 5, 1894, p. 246; Jan. 3, 1896, p. 319; Jan. 5, 1900, p. 370.

35. *Electrical World,* Jan. 3, 1885, p. 1.

36. "The Convention of the National Electric Light Association at New York, May 5–7, 1896," *Western Electrician* (Chicago), May 16, 1896, p. 239.

37. "Exhibits of Electrical Interest at the Mechanic Fair, Boston," *Electrical World,* Oct. 29, 1898, p. 451.

38. "Crowns of Living Light," *Electrical World,* Dec. 6, 1884, p. 235.

39. "Exposition Lighting," *Electricity* (London), June 28, 1899, pp. 392–93.

40. *Electrical Review,* June 23, 1888, p. 7.

41. "A Novel Electric Tower," ibid., Aug. 11, 1888, p. 10.

42. J. P. Barrett, "Electrical Industries and the World's Fair," *Electrical World,* Aug. 30, 1890, p. 2.

43. "A Novel Entertainment," *Electrical Review,* Dec. 17, 1892, p. 205.

44. "The Electric Tower at the Minneapolis Exposition," *Western Electrician* (Chicago), Sept. 20, 1890, p. 151.

45. Ibid.

46. "The Illumination of Fountains by Electricity," *Electrical World,* Nov. 8, 1884, p. 177.

47. "The International Electrical Exhibition," *Electrical World,* Aug. 30, 1884, p. 65; "The Illumination of Fountains."

48. *Western Electrician* (Chicago), June 20, 1896, p. 306.

49. "Electrical Features of the Dedication of the World's Fair in Chicago," *Western Electrician* (Chicago), Oct. 29, 1892, p. 217.

50. "How To Produce the Effect of a Waving Flag," *American Electrician,* Apr. 1898, pp. 147–48.

51. "Reading at Seven and a Half Miles Distance from the Candle," *Scientific American,* Sept. 6, 1879, p. 145.

52. "The Electric Light," *Electrical World,* Apr. 12, 1884, p. 123.

53. "Electrical Features of the Dedication of the World's Fair," p. 217.

54. "The Washington Illumination," *Western Electrician* (Chicago), Oct. 5, 1892, p. 187.

55. "The Inaugural Ball Illumination," *Electrical Review,* March 16, 1889,

p. 9; "Electrical Features of the Bicycle Show," *Western Electrician* (Chicago), Jan. 18, 1896, p. 1.

56. "Electrical Illumination at the Peace Jubilee," *Western Electrician* (Chicago), Oct. 29, 1898, p. 39.

57. "The Electrical Display During the Columbian Celebration," *Electrical World*, Oct. 22, 1892, p. 256.

58. "Electrical Decorations at Detroit," *Western Electrician* (Chicago), Sept. 12, 1892, p. 153.

59. "The Washington Illumination," *Western Electrician* (Chicago), Oct. 8, 1892, p. 187.

60. Ibid.

61. "Electrical Features of the Dedication of the World's Fair," p. 217.

62. "Electrical Decorations in London," *Western Electrician* (Chicago), June 19, 1897, p. 359. Quoted from "an English contemporary," which in its turn quotes from the *Daily Telegraph*.

63. *Comfort* (Augusta, Maine), Dec. 1893, p. 8.

64. Trumbull White and William Igelheart, *The World's Columbian Exposition. Chicago, 1893* (Philadelphia: International Publishing Co., 1893), p. 302.

65. Ibid., pp. 322–23; Ben C. Truman, *History of the World's Fair, Being a Complete Description of the World's Columbian Exposition From Its Inception* (Chicago: Cram Standard Book Co., 1893), pp. 358–59; *A Week at the Fair: Illustrating the Exhibits and Wonders of the World's Columbian Exposition* (Chicago: Rand, McNally, 1893), p. 79.

66. *A Week at the Fair,* ibid.

67. Ibid.

68. Ibid., p. 73.

69. Rossiter Johnson, ed., *A History of the World's Columbian Exposition,* 4 vols. (New York: D. Appleton, 1897), 1:510.

70. *Electrical World,* Jan. 17, 1885, p. 21.

71. "Turning Off the Gas in Paris," *Electrical Review,* Sept. 18, 1886, p. 4.

72. "Trolley Parties as Viewed by Steam Railway Men," *Western Electrician* (Chicago), Aug. 8, 1896, p. 67, quoted from *Railway Age;* "Electrical Trolley Riders," *Western Electrician* (Chicago), Aug. 19, 1899, p. 107; "By the Bright Light," *Electrical World,* July 19, 1884, p. 4.

73. "Pleasure Travel on Electric Railways," *Western Electrician* (Chicago), Aug. 29, 1896, p. 99, quoted from *Railway Age.*

74. Ibid.

75. *Electrical Review,* Sept. 1, 1888, p. 4.

76. Ibid., July 10, 1886, p. 9.

77. Remarks by William Preece in a speech before the Society of Arts in London, quoted in "The Light That Will Extinguish Gas," *Electrical Review,* Feb. 7, 1885, p. 2.

78. "Electric Lighting Effects at the Madison Square Garden," *Electrical Review*, Nov. 14, 1891, p. 169.

79. Glenn Marston, "Light To Advertise Cities," *Technical World Magazine* (Chicago), Oct. 1911, p. 201.

80. "Mr. Edison as an Electrical Litterateur."

81. "Electric Lighting in Theatres," *Electrical World*, May 10, 1884, p. 152. Quoted from *La Lumière Électrique*.

82. "The Opera-House, Paris," *Electrical World*, July 19, 1884, p. 19.

83. Beane, *Proceedings*, p. 32.

84. "Electric Lights in the Fifth Avenue Theatre," *Electrical World*, Sept. 6, 1890, p. 173.

85. "The Electric Light," *Electrical World*, Apr. 12, 1884, p. 123.

86. "Electric Lighting on the Stage," *Electrician* (London), Nov. 25, 1882, p. 27.

87. "Electric Lighting Effects in the Ballet," *Western Electrician* (Chicago), Dec. 26, 1896, p. 315.

88. "The Electric Light for Santa Claus," *Electrical World*, Jan. 12, 1884, p. 884.

89. "A Christmas Tree Lighted by Electricity," *Electrical World*, Jan. 3, 1885, p. 6.

90. "Setting Off Fireworks by Electricity," *Electrical World*, July 19, 1890, pp. 43–44.

91. "Electric Installation at Greenwich, Ct.," *Electrical Review*, Oct. 22, 1887, p. 1.

92. "Electricity Furnished by the Cartload," *Electrical Review*, Jan. 28, 1888, p. 1.

93. "Electrical Spectacular Effects," *Western Electrician* (Chicago), Apr. 8, 1899, p. 196.

94. *Electrical Review*, Mar. 19, 1887, p. 5.

95. See, for example, "Electric Night Signals," *Electrical Review*, June 23, 1888, p. 1, for an account of using electric signal lights in place of semaphores, and "Miscellaneous Notes," *Western Electrician* (Chicago), May 21, 1898, p. 297.

96. George H. Drayer, "Electricity at the National Capitol," *Electrical Review*, Dec. 18, 1895, p. 345.

97. William T. O'Dea, *The Social History of Lighting* (London: Routledge and Kegan Paul, 1958), p. 178.

98. "An Electrical Wedding," *Electrical Review*, Nov. 12, 1892, p. 138.

99. "A Wedding with Electrical Accessories," *Western Electrician* (Chicago), Dec. 30, 1899, p. 381.

100. "Electrically Illuminated Signs," *Western Electrician* (Chicago), July 16, 1892, p. 39.

101. "Advertising—A Mode of Motion," *Scientific American*, May 18,

1878, p. 306. See also "Blackmer's Electrical Sign," *Telegraphic Journal* (London), Apr. 15, 1879, p. 133.

102. *Western Electrician* (Chicago), May 16, 1896, p. 246.

103. "Smart Advertising Booms," *Answers* (London), Aug. 2, 1890, p. 150.

104. *Electrical Engineer* (London), Jan. 4, 1895, p. 18.

105. "Electrical Advertising," ibid., Apr. 26, 1895, p. 465. See also *Invention* (London), Jan. 22, 1898, p. 51.

106. "Views, News and Interviews," *Electrical Review*, Apr. 11, 1891, p. 107.

107. Quoted in "Electrical Decorations in London," p. 359.

108. *Electrical Review*, Sept. 3, 1892, p. 13. Quoted from "Meteor" in *Lightning*.

109. "Electrical Decorations in London."

110. "Electricity and the Birthday of the German Emperor," *Electrical Review*, May 14, 1887, p. 3.

111. "Electrical Decorations in New York," *Western Electrician* (Chicago), Oct. 14, 1899, p. 217.

112. "Electrical Illumination at the Chicago Festival," *Western Electrician* (Chicago), Oct. 14, 1899, p. 217.

113. "Christmas and Electricity," *Electrical World*, Dec. 22, 1894, p. 639.

114. "Cloud Telegraphy," *Electrical Review*, May 5, 1888, p. 5.

115. "The Electric Light as a Military Signal," *Scientific American*, Oct. 30, 1875, p. 281.

116. *Electrical Review*, Oct. 29, 1887, p. 9.

117. Ibid., Jan. 7, 1888, p. 7.

118. "Advertising in the Clouds," *Electrical Review*, Oct. 6, 1888, p. 4.

119. "Advertising in the Clouds: Its Practicability," *Electrical World*, Dec. 31, 1892, p. 427.

120. "People Who Suggest Advertising Ideas," *Answers* (London), Mar. 26, 1892, p. 310.

121. *Electrical Review*, Dec. 21, 1889, p. 4.

122. "Even the Clouds Don't Escape Him," *Electrical World*, Nov. 26, 1892, p. 335.

123. "Advertising in the Clouds."

124. "Future Electrical Development," interview with Amos Dolbear, *Western Electrician* (Chicago), Jan. 9, 1897, p. 24.

125. "Advertising on the Clouds," *Invention* (London), Feb. 17, 1894, pp. 150–51.

126. "Future Electrical Development."

127. *Electrical Review*, Nov. 12, 1892, p. 137.

128. "Advertising on the Clouds."

129. "Announcing Election Returns by Search-Light Signals," *Electrical Review*, Nov. 14, 1891, p. 166.

130. "Election Returns," *Electrical Review*, Nov. 19, 1892, p. 151.

131. Ibid.

132. "The Newest Horror," *Answers* (London), July 16, 1892, p. 129.

133. Quoted in "Even the Clouds Don't Escape Him."

134. "A Message from the Moon," *Science Siftings* (London), Nov. 16, 1895, p. 77.

135. "Future Electrical Development."

136. Amos Dolbear, "The Electric Searchlight," *Cosmopolitan*, Dec. 1893, p. 254.

137. Arthur Bennington, "Some of the Plans of Science To Communicate with Mars. 40,000,000 Miles Away in the Depths of Infinite Space," *Live Wire*, Feb. 1908, p. 6.

138. Ibid.

139. Ibid.

140. "Tesla's Signals from Mars," *Popular Science News* (Boston), Mar. 1901, p. 49.

141. "The Prophet's Column," *Electrical Review*, Oct. 3, 1891, p. 78.

142. Bennington, "Some of the Plans."

143. *Spectator* (London), Nov. 26, 1892, p. 765.

5   Annihilating Space, Time, and Difference

1. Amos Dolbear, "Electricity and Civilization," manuscript, Amos Dolbear papers, Tufts University, p. 19. The manuscript is undated, but mentions Marconi and says that telegraphy has been in use for sixty years, which dates it at about 1900.

2. Alonzo Jackman, letter to the editor of the *Woodstock* (Vermont) *Mercury*, Aug. 14, 1846, in "Some of the Early History of Telegraph Cable Manufacture," *Electrical Review*, Nov. 23, 1889, p. 2.

3. Nikola Tesla, "Experiments with Alternating Current of High Potential and High Frequency," 1904, in Edwin H. Armstrong, "The Progress of Science: Nikola Tesla, 1857–1943," *Scientific Monthly* (Washington, D.C.), Apr. 1943, pp. 379–80.

4. Nikola Tesla, "The Problem of Increasing Human Energy," *Century Magazine*, June 1900, p. 183.

5. "Electrical Possibilities as Viewed by a Business Man," *Electrical Review*, June 15, 1889, p. 2.

6. "Looking Forward," *Science Siftings* (London), May 6, 1893, pp. 40–41.

7. *All the Year Round* (London), Oct. 6, 1894, p. 328.

8. "A Possible Marvel," *Electrical Review*, May 23, 1891, p. 172.

9. "Telegraphy Without Wires," *Electrician* (London), June 21, 1895, p. 250, quoting Amos Dolbear in *Electrical Engineer*, May 29, 1895.

10. J. W. Wilkins, letter to the editor, Mar. 28, 1849, *Mining Journal and Railway Gazette* (Boston), Mar. 31, 1849, pp. 157–58.

11. Amos Dolbear, "Electricity and Civilization." See also Dolbear, "The Future of Electricity," *Donahoe's Magazine* (Boston), Mar. 1893, pp. 291–92.

12. "From an Admirer of the Telephone," *Electrical Review*, Nov. 23, 1889, p. 6.

13. *Electrical Review*, May 25, 1889, p. 6.

14. W. H. Preece, "Electric Signalling Without Wires," speech to the Society of Arts, reported in "The Transmission of the Electric Current Through Space Without Wires," *Invention* (London), Mar. 3, 1894, pp. 188–89.

15. *The Million* (London), Aug. 27, 1892, p. 230.

16. Arthur Bennington, "Some of the Plans of Science To Communicate with Mars," *Live Wire*, Feb. 1908, p. 8.

17. Thomas Stevens, "Telegraph Operators in Persia," *Electrical Review*, Aug. 18, 1888, p. 6.

18. *Electrical Review*, Jan. 3, 1891, p. 227.

19. Ibid., Oct. 3, 1885, p. 7.

20. "The Moral Influence of the Telegraph," *Scientific American*, Oct. 15, 1881, p. 240.

21. "The Telectroscope," *Invention* (London), Apr. 23, 1898, p. 267.

22. "The Telephote," *Electrical Review*, Aug. 24, 1889, p. 120. See also "The Industrial Development [sic] of Electricity," *All the Year Round* (London), Oct. 6, 1894, p. 330: "Professor Bell is convinced that in the near future it will be possible to see by telegraph, so that a couple conversing by telephone can at the same time see each other's faces. Extending the idea, photographs may yet be transmitted by electricity, and if photographs, why not landscape views? Then the stay-at-home can have the whole world brought before his eyes in a panorama without moving from his chair."

23. Edward Bellamy, *Equality* (New York: D. Appleton, 1897), pp. 347–48.

24. "Sight Transmission of Pictures," *Comfort* (Augusta, Maine), Apr. 1898, p. 20.

25. Alonzo Jackman, letter to the editor, *Woodstock Mercury*, Aug. 14, 1846.

26. Julian Hawthorne, "June, 1993," *Cosmopolitan*, Feb. 1893, p. 456.

27. Ibid., p. 457.

28. Ibid., p. 456.

29. Ibid.

30. Elizabeth Eisenstein, *The Printing Press as an Agent of Change: Transformations in Early-Modern Europe* (New York: Cambridge University Press, 1979), vol. 1.

31. "The Extension of Sense," *Scientific American*, Oct. 16, 1876, p. 384.

32. "The Good the Speech Recorder Might Have Done," *Electrical Review*, Aug. 25, 1888, p. 7. See also *Science Siftings* (London), on the possibility of Edison's kinetoscope in the House of Commons reflecting the eloquence of Mr. Gladstone, "Interview with Edison. The Kinetoscope," Apr. 28, 1894, pp. 24–25: "What we would not give now at the present day to be able to have before us not only the portraits but the very actions and speech of the great men of the past."

33. "The Phonograph and Music," *Electrical Review*, May 26, 1888, p. 8. Quoted from the *New York Evening Post*.

34. "The Future of the Phonograph," *Electrical Review*, Apr. 21, 1888, p. 1.

35. *Chambers's Journal* (London), Feb. 28, 1891, p. 142.

36. "The Phonograph as a Ruler of Posterity," quoted in *Electrical World*, Aug. 4, 1888, p. 55.

37. *Lightning* (London), Dec. 24, 1896, p. 513.

38. The text of Mendenhall's address was printed in *Popular Science Monthly*, Nov. 1890, pp. 19–31.

39. "The Convention of the National Electric Light Association," *Western Electrician* (Chicago), May 16, 1896, p. 238; remarks of May 5.

40. "What Did George the Third Know?", *Science Siftings* (London), Feb. 10, 1894, p. 230.

41. E. B. Dunn, "The Storing of Atmospheric Electricity," *North American Review* (Boston), July 1897, p. 123.

42. G. H. Knight, "Relation of Invention to the Conditions of Life," *Cosmopolitan*, Feb. 1892, p. 44.

43. "The Electrical Firmament," *Electrical World*, Mar. 3, 1894, p. 273.

44. "The Electrical Exhibition at Paris," *Electrician* (London), Dec. 3, 1881; "The International Exhibition and the Congress of Electricity at Paris," *Scientific American*, Dec. 10, 1881, p. 377.

45. Edouard Hospitalier, *Modern Applications of Electricity*, Julius Maier, trans. (London: Kegan Paul, 1882), p. 392.

46. *Telegraphic Journal and Electrical Review* (London), Dec. 1, 1881, pp. 485–86.

47. "Opera by Telephone," *Scientific American*, June 14, 1884, p. 373.

48. "Music by Telephone in Germany," *Electrical World*, May 3, 1884, p. 144.

49. "A New Feat in Telephony," *Electrical World*, Sept. 27, 1884, p. 108.

50. "Theatrophone Service in Paris," *Western Electrician* (Chicago), May 14, 1892, p. 288. See also "Theatrophones in Paris," *The Telephone* (Chicago), Sept. 2, 1889, p. 406.

51. *Western Electrician* (Chicago), Sept. 12, 1891, p. 155.

52. *Invention* (London), Feb. 1, 1896, p. 66.

53. *Lightning* (London), July 9, 1896, p. 38.

54. "Telephone Amusement," *Electrical Review,* Dec. 7, 1889, p. 7. Reprinted from the *London Globe.*

55. *Electrician* (London), Dec. 11, 1891, p. 137.

56. "The Modern Wonder-Worker, Electricity," *Chambers's Journal* (London), Mar. 19, 1892, p. 178.

57. "Telephones in Churches," *Lightning* (London), Nov. 19, 1896, p. 436.

58. *Electrical Review,* July 9, 1887, p. 7.

59. "Applause by Telephone," ibid., Jan. 7, 1888, p. 5.

60. "Among the 'Phones," *Scientific American,* July 12, 1890, p. 21. Quoting from *Modern Light and Heat.*

61. "The Recent Telephone-Phonograph Experiments," *Electrical Review,* Feb. 16, 1889, p. 4.

62. *Electrical Review,* Aug. 27, 1887, p. 6.

63. "The Telephone," *Western Electrician* (Chicago), Sept. 19, 1890, p. 39.

64. "Foreign Currents," *American Telephone Journal* (Chicago), Apr. 26, 1902, p. 280; *Electrical Review,* June 2, 1897, p. 261.

65. "Concert Music by Telephone," *Scientific American,* Oct. 10, 1891, p. 225.

66. "A Telephone Musicale," *Telephony* (Chicago), July 1905, p. 52.

67. "Phonographic Music Transmitted by Telephone," *Electrical Review and Western Electrician,* Sept. 12, 1912, p. 567.

68. "Comic Opera by Telephone," *Western Electrician* (Chicago), Mar. 4, 1899, p. 126.

69. "Telephone Concerts in Wisconsin," *Western Electrician* (Chicago), Dec. 26, 1896, p. 315.

70. Ibid.

71. "The Telephone as Entertainer," *Telephony* (Chicago), May 18, 1912, p. 610.

72. "Base Ball by Electricity," *Electrical Review,* July 24, 1886.

73. Ibid. Quoted from the *Detroit Free Press.*

74. "Long Distance Telephoning," *Electrical Review,* Aug. 3, 1889, p. 6.

75. *Electrical Review,* June 15, 1889, p. 4.

76. "Baseball by Electricity," *Electrical World,* June 9, 1894, p. 778.

77. "Tapping the Telegraph Wires," *Science Siftings* (London), June 4, 1892, p. 87.

78. W. J. Baker, *A History of the Marconi Company* (London: Methuen, 1970), p. 39.

79. *Western Electrician* (Chicago), Sept. 9, 1898, p. 147.

80. Baker, *A History of the Marconi Company,* p. 486; George H. Perry, "Simple Explanations: Wireless Telegraph," *Everybody's Magazine,* Dec. 1899. There were, of course, telegraphic precedents. See the *Telegraphic Journal*

(London), Apr. 1, 1875, pp. 68–80, which describes telegraphic reporting of the Oxford-Cambridge Boat Race from Putney to Morlake in 1872.

81. *Electrical World*, 5:1. See also "Church Service by Telephone," *American Telephone Journal* (Chicago), July 30, 1904, pp. 65–66.

82. "The Telephone and Revival Meetings," *Telephony* (Chicago), May 25, 1912, p. 644.

83. "The Telephone at Christ Church, Birmingham, England," *Western Electrician* (Chicago), Oct. 11, 1890, p. 195. See also *Electrical Review* (London), 14(17):7.

84. "Looking Backward," *Electrician* (London), Feb. 11, 1892, p. 366; "The Electrophone," *Electrical Engineer* (London), July 19, 1895, p. 195.

85. "A Stump Speech over the Telephone," *Western Electrician* (Chicago), Sept. 12, 1896, p. 234.

86. "Extraordinary Telephoning," *Tit-Bits* (London), Jan. 9, 1897, p. 280. The most interesting account of this event appears in "Electrical Features of the Chicago Day Celebration," *Western Electrician* (Chicago), Oct. 17, 1896, pp. 182–84.

87. "Novel Use of the Telephone," *Western Electrician* (Chicago), Aug. 16, 1890, p. 185.

88. "How Electricity Will Help Out the Editor of the Future," *Electrical Review*, Feb. 2, 1889, p. 4. Verne's story is reprinted in Jules Verne, *Yesterday and Tomorrow*, I. O. Evans, trans. (Westport, Conn.: Associated Booksellers, 1965), pp. 107–24.

89. "The Industrial Development of Electricity."

90. *Electrical Review*, Sept. 1, 1888, p. 10.

91. "Wireless as a News Carrier," *Electrical Review and Western Electrician*, Apr. 6, 1912, p. 668.

92. *Electrical Review*, Nov. 17, 1888, p. 5.

93. "Election Returns by Telephone and Telegraph," *Western Electrician* (Chicago), Nov. 26, 1892, p. 275; "Election Returns," *Electrical Review*, Nov. 19, 1892, p. 151.

94. "Election Returns," *Electrical Review*, Nov. 19, 1892, p. 151.

95. "The Telephone on Election Day," *Electrical Review*, Nov. 11, 1896, p. 237.

96. Ibid.

97. "Use of Electricity in Announcing Election Returns," *Western Electrician* (Chicago), Nov. 17, 1900, p. 323.

98. Ibid.

99. M. D. Atwater, "Distributing National Election Returns by Telephone," *Telephony* (Chicago), Nov. 9, 1912, pp. 721–24.

100. "The Telephone and Election Returns," *Electrical Review*, Dec. 16, 1896, p. 298.

101. Ibid.

102. *Electrical Engineer* (London), Sept. 4, 1895, p. 233.

103. Ibid., Aug. 28, 1895, pp. 206–207.

104. "Election Night in New York," *Harper's Weekly*, Nov. 14, 1896, p. 1122.

105. "The Farmer and the Telephone," *American Telephone Journal* (Chicago), Mar. 29, 1902, pp. 202–203.

106. "War News in Chicago by Telephone," *Western Electrician* (Chicago), Apr. 9, 1898, p. 213.

107. "The Telephone in the Country," *Telephony* (Chicago), July 1905, p. 52.

108. "Telephoning is the National Craze," *Telephony* (Chicago), Dec. 1905, p. 52.

109. *Lightning* (London), Jan. 5, 1893, p. 1.

110. Paul Adorjan. "Wire-Broadcasting," *Journal of the Society of Arts* (London), Aug. 31, 1945, p. 514. See also *Electrical Engineer* (London), July 19, 1895, p. 57, and *Electrician* (London), June 9, 1899, p. 243.

111. "The Telephone in Hungary," *Scientific American*, July 2, 1881, p. 5.

112. John J. O'Neill, *Prodigal Genius. The Life of Nikola Tesla* (New York: Ives Washburn, 1944), pp. 45–46.

113. "The Telephone Journal," *Science Siftings* (London), July 15, 1893, p. 362.

114. Ferenc Erdei, ed., "Radio and Television," *Information Hungary* (New York: Pergamon Press, 1968), p. 645; "Telephon-Zeitung," *Zeitschrift für Elektrotechnik* (Vienna), Dec. 1, 1896, p. 741.

115. Thomas S. Denison, "The Telephone Newspaper," *World's Work*, Apr. 1901, p. 641.

116. W. G. Fitzgerald, "A Telephone Newspaper," *Scientific American*, June 22, 1907, p. 507; Frederick A. Talbot, "A Telephone Newspaper," *Littell's Living Age* (Boston), Aug. 8, 1903, pp. 374–75.

117. Denison, "The Telephone Newspaper," p. 642.

118. "Telephon-Zeitung," *Zeitschrift für Elektrotechnik* (Vienna), p. 741.

119. "The Telephone Newspaper," *Electrical Engineer* (London), Sept. 6, 1895, p. 257.

120. Toth Endréné, ed., *Budapest Enciklopédia* (Budapest: Corvina Kiadó, 1970), p. 313.

121. Fitzgerald, "A Telephone Newspaper," p. 507; Talbot, "A Telephone Newspaper," p. 375.

122. "Telephon-Zeitung," *Zeitschrift für Elektrotechnik* (Vienna), pp. 740–41.

123. Talbot, "A Telephone Newspaper," p. 507. By 1907 the list of foreign exchanges included New York, Frankfurt, Paris, Berlin, and London.

124. Denison, "The Telephone Newspaper," p. 642.

125. *A Pallas Nagy Lexicon* (Budapest: Pallas Irodalmi, 1897), 16:20–21.

126. "The Telephone Newspaper," *Electrical Engineer* (London), p. 257; Fitzgerald, "A Telephone Newspaper," p. 507.

127. Jules Erdoess, "Le Journal Téléphonique de Budapest: L'Ancêtre de la Radio," *Radiodiffusion* (Geneva), Oct. 1936, p. 37.

128. W. B. Forster Bovill, *Hungary and the Hungarians* (London: Methuen, 1908), p. 111.

129. "A Talking Newspaper," *Invention* (London), Mar. 26, 1898, p. 203; Talbot, "A Telephone Newspaper," p. 375.

130. Denison, "The Telephone Newspaper," p. 642.

131. See, for example, *Lightning* (London), Feb. 23, 1893, p. 115; "A Telephone Newspaper," *Newspaper Owner and Manager*, May 4, 1898, p. 22; "The Telephone Newspaper," *Scientific American,* Oct. 26, 1895, p. 26, reprinted from the *New York Sun*.

132. Erdoess, "Le Journal Téléphonique de Budapest," p. 39.

133. Arthur F. Colton, "Telephone Newspaper—A New Marvel," *Technical World Magazine* (Chicago), Feb. 1912, p. 668.

134. "Order in the Matter of the Petition of the New Jersey Herald Telephone Company," *Second Annual Report of the Board of Public Utility Commissioners for the State of New Jersey* (Trenton: MacCrellish and Quigley, 1912), pp. 147–50.

135. Colton, "Telephone Newspaper," p. 669.

136. Ibid.

137. Ibid.

138. Erik Barnouw, *A History of Broadcasting in the United States,* 3 vols. *A Tower in Babel*, vol. 1 (New York: Oxford University Press, 1969), p. 100.

139. Colton, "Telephone Newspaper," p. 669.

140. "An American Telephone Newspaper," *Literary Digest,* Mar. 16, 1912, p. 529, quoting *Editor & Publisher*.

141. " 'Phone Herald's Short Life," *Fourth Estate*, Mar. 2, 1912, p. 23.

142. William Peck Banning, *Commercial Broadcasting Pioneer. The WEAF Experiment, 1922–1926* (Cambridge: Harvard University Press, 1946), p. 59.

143. Ibid., p. 60.

Index